COMPETING WITH THE SOVIETS

Michael ~

    You've been an inspiration
since my earliest days of
grad school. Thanks so
much for all your goodwill
and encouragement over the
last 15 years!

        Audra

JOHNS HOPKINS
INTRODUCTORY STUDIES
IN THE HISTORY
OF SCIENCE

Mott T. Greene
and Sharon Kingsland
Series Editors

Competing with the Soviets

Science, Technology, and the State in Cold War America

Audra J. Wolfe

THE JOHNS HOPKINS UNIVERSITY PRESS

BALTIMORE

© 2013 The Johns Hopkins University Press
All rights reserved. Published 2013
Printed in the United States of America on acid-free paper
9 8 7 6 5 4 3 2 1

The Johns Hopkins University Press
2715 North Charles Street
Baltimore, Maryland 21218-4363
www.press.jhu.edu

Library of Congress Cataloging-in-Publication Data

Wolfe, Audra J.
    Competing with the Soviets : science, technology, and the state in Cold War
America / Audra J. Wolfe.
        p. cm. — (Johns Hopkins introductory series in the history of science)
    Includes bibliographical references and index.
    ISBN 978-1-4214-0769-2 (hdbk. : alk. paper) — ISBN 978-1-4214-0771-5
(pbk. : alk. paper) — ISBN 1-4214-0769-8 (hdbk. : alk. paper) — ISBN 1-4214-
0771-X (pbk. : alk. paper)
    1. Science—United States—History—20th century. 2. Science—Soviet
Union—History—20th century. 3. Technology—United States—History—
20th century. 4. Technology—Soviet Union—History—20th century. 5. World
politics—1945–1989. 6. Cold War. I. Title.
    Q127.U6W65 2013
    509.73'09045—dc23   2012012930

A catalog record for this book is available from the British Library.

The illustration on page 51 is from Derek J. de Solla Price, *Science Since Babylon*
(New Haven: Yale University Press, 1961).

The illustration on page 109 is from the Ava Helen and Linus Pauling Papers,
Oregon State University Libraries Special Collections.

*Special discounts are available for bulk purchases of this book. For more information,
please contact Special Sales at 410-516-6936 or specialsales@press.jhu.edu.*

The Johns Hopkins University Press uses environmentally friendly book materials, including recycled text paper that is composed of at least 30 percent post-
consumer waste, whenever possible.

Contents

Abbreviations

AAAS	American Association for the Advancement of Science
ABM	anti-ballistic missile
AEC	Atomic Energy Commission
APL	Applied Physics Laboratory (Johns Hopkins University)
ARPA	Advanced Research Projects Agency (U.S. Department of Defense)
CENIS	Center for International Studies (Massachusetts Institute of Technology)
CIA	Central Intelligence Agency
CNI	Committee for Nuclear Information
CSM	command/service module
DOD	U.S. Department of Defense
DOE	U.S. Department of Energy
ELDO	European Launcher Development Organization
FAO	Food and Agriculture Organization of the United Nations
FAS	Federation of American Scientists
GAC	General Advisory Committee (Atomic Energy Commission)
ICBM	intercontinental ballistic missile
IGY	International Geophysical Year
IIT	Indian Institute of Technology
IRBM	intermediate-range ballistic missile
LEM	lunar excursion module
MIT	Massachusetts Institute of Technology
NAS	National Academy of Sciences
NASA	National Aeronautics and Space Administration
NIH	National Institutes of Health
NSF	National Science Foundation
OEO	Office of Economic Opportunity
ONR	Office of Naval Research
OSRD	Office of Scientific Research and Development

PPBS   Planning, Programming, and Budgeting System
PSAC   President's Science Advisory Committee
R&D   research and development
RPP&E   Division of Research, Plans, Programs, and Evaluation (Office of
          Economic Opportunity)
SAGE   semi-automatic ground environment
SDC   System Development Corporation
SDI   Strategic Defense Initiative
TVA   Tennessee Valley Authority
UN   United Nations
UNESCO   United Nations Educational, Scientific, and Cultural
          Organization
USAID   U.S. Agency for International Development

COMPETING WITH THE SOVIETS

Introduction

"If anyone wants a hole in the ground," physicist Edward Teller is reputed to have said in 1962, "nuclear explosives can make big holes."* Teller's enthusiasm for what he called "nuclear engineering" drove Project Plowshare, an ill-fated attempt to use atomic weaponry for peaceful purposes. Between 1957 and 1975, Teller and his colleagues spent hundreds of millions of dollars devising plans to use nuclear devices as convenient tools for mining operations, oil and gas exploration, and most famously, earthmoving projects. Bombs might be used to create a new Alaska harbor or, perhaps, a new Panama Canal. Project Plowshare's advocates believed that such endeavors could be completed safely, without excessive danger to persons or the environment, and that successful nuclear construction projects might ultimately save the government a small fortune on critical infrastructure.

More than fifty years after it was initially proposed, Project Plowshare can be read as a symbol of everything that was wrong with science and technology in the Cold War. It assumed that civilian applications would follow naturally from military research. Most of its reports were classified. Its continued survival depended on the support of powerful political and scientific sponsors who were infatuated with atomic physics and obsessed with nuclear weaponry. University scientists who criticized the program's goals and assumptions lost their jobs. Its technological hubris, at least in retrospect, seems more than mildly ridiculous. And like so many other Cold War technological projects, it left environmental destruction in its wake.

But, like the Cold War itself, the real story of Project Plowshare must be told through shades of gray, not black and white. Its high-tech approach made sense at a time when science—particularly atomic science—seemed to offer the best solutions to the nation's problems, whether those problems might involve infrastructure or foreign policy. Nor did all of Project Plowshare's research have military ends: work on the biological effects of radiation underwritten

*Edward Teller, as quoted in Scott Kirsch, *Proving Grounds: Project Plowshare and the Unrealized Dream of Nuclear Earthmoving* (New Brunswick, NJ: Rutgers University Press, 2005), p. xiv.

by Plowshare funds helped establish the field of systems ecology. Support for the project was neither inevitable nor unanimous; it encountered resistance from scientists, journalists, environmentalists, politicians, and Native peoples at every step of the way. Project Plowshare does indeed perfectly capture the spirit of Cold War science, but not necessarily as an exercise in technocracy run amok. Rather, it serves as a vivid illustration of both the faith that postwar Americans placed in state-sponsored science and the doubts that simmered just below the surface of this consensus.

This book is an attempt to tell the story of science and technology, particularly American science and technology, during the Cold War. As treated here, the "Cold War" refers to the period of intense conflict between the United States, the Soviet Union, and their associated allies from the end of World War II to the collapse of the Soviet Union in 1991. Intended as an introductory synthesis rather than as a comprehensive account, it uses certain key episodes, anecdotes, and individuals to show the central—and unique—position that science and technology held in relation to the Cold War state.

Some of the key themes that emerge from these stories will be familiar to students of American history; others are specific to the history of science. At the more general level, the science-state relationship as it developed over the course of the Cold War reflected broader attitudes about the proper role of government in American life. The period between the start of the New Deal and the early 1970s marked the largest expansion of government in the history of the United States. Scientific research, like so many other areas of American life, subsequently fell under the umbrella of federal support, and thereby, federal oversight. As government programs began to fall out of favor with political commentators and conservative voters, so too did unlimited support for research in science and technology. Similarly, the shift from small-scale, individualist research practices to team endeavors in large, multidisciplinary, hierarchical laboratories mirrors a familiar theme from American business history. In the realm of social history, scientists were subjected to the same kinds of security hearings and loyalty investigations as other American citizens in the 1950s, and at least some of them participated in the anti–Vietnam War protests so common on university campuses in the 1960s.

Yet other themes, specific to the history of science, can deepen our understanding of postwar American history. The special role of science and technology in constructing the atomic bomb, that terrifying symbol of the 1950s, seemed to create an equally special role for scientific experts as figures of political authority. The question of the proper role of a technocratic elite figured prominently in policy circles both at home and abroad, as officials at institutions as varied as city governments and the U.S. Department of State struggled with integrating scientific expertise into their decision-making processes. Similarly, an understanding of how American science came to stand in for democracy in the minds of some policymakers is critical for making sense of much of 1960s-era foreign policy.

And, of course, the Cold War changed the practice of science itself. Attempts to understand the character and effects of science in the Cold War reached critical mass when President Dwight D. Eisenhower used his 1961 farewell address to warn the nation of the creeping influence of what he called "the military-industrial complex." By this Eisenhower was referring to the explosive growth of military research, development, and production, almost all of which was handled by outside industrial or academic contractors. Because so much of postwar military R&D relied on cutting-edge science, the rise of the military-industrial complex had been accompanied by the creation of elaborate, though uncoordinated, structures for federal science advising. Eisenhower's remarks gave name to a phenomenon that critics, including scientists, had been watching with growing dismay since the end of World War II. By 1961, the nation's aggressive investment in military and scientific research had yielded a fully stocked nuclear arsenal that promised, instead of peace and prosperity, mutually assured destruction.

It is perhaps in part because of Eisenhower's criticism that so much of historians' initial work on Cold War science focused on the questions of whether, or how, military funding was distorting the practice and culture of science. Studies have explored how postwar scientists chose their research topics, organized their research teams, used their instruments, sought patrons, and dealt with security concerns. While historians generally agree that the military-industrial complex exists and that science and technology were key components of it, they disagree over who was using whom. To give a classic example: most of the particle accelerators currently in place in the United States were purchased with federal support with the implicit assumption that the findings and expertise the equipment produced might some day prove useful in developing weaponry. Scientists who worked at such facilities were subject to security restrictions and were generally hesitant to do anything that might upset their patrons at the Atomic Energy Commission (AEC). At the same time, these devices made it possible for scientists to discover new chemical elements and subatomic particles—discoveries that had few, if any, military applications. Defense and civilian needs dovetailed in the complicated and contested world of Cold War science.

This book largely skirts the question of scientific "distortion." That term

implies that there exists some sort of pure, undistorted science devoid of political or cultural intention. A generation's worth of scholarship in the history
of science has demonstrated that, to the contrary, scientists are participants in
the culture in which they live. Their choices and opportunities have always
been shaped by the ideological assumptions, political mandates, and social
mores of their times. Indeed, the idea that American science ever operated in
a "free zone" outside of politics is itself a legacy of the ideological Cold War.
In the postwar United States, the vast majority of scientists assumed that the
federal government had an interest in promoting scientific research and that,
ultimately, this research would produce a better, stronger society. There were,
of course, those who disagreed with this line of thinking, and we shall return
to their stories toward the end of this book.

The question of how the Cold War affected the practice of science has a
flip side: how did the state use science? Eisenhower's comments are most likely
to bring to mind images of missiles, fighter jets, and atomic bombs, a trio
that quite rightly points to the centrality of high-tech weaponry to national
security. Foreign policy, however, is about more than just warfare. Consider,
for example, the collaborative attempts of American and Indian agricultural
scientists to develop more productive and nutritious varieties of wheat. It was
hoped that this project, supported with funds from the Department of State,
the Department of Agriculture, and American seed companies, might prevent
starvation and accompanying unrest among the exploding population on the
Indian subcontinent, thereby giving support to a fragile democracy that the
United States hoped to hold in its orbit. Moreover, those involved with the
project believed that exposing Indian policymakers to American agricultural
products might open new markets. From the Indian perspective, partnerships
with American development experts offered the quickest route to modernization and, they hoped, economic independence. The products of science could
therefore offer multiple foreign policy rewards, from promoting goodwill and
developing alliances to ensuring economic dominance. Science could be a car-
rot as well as a stick.

Postwar American policymakers believed in the power of science to solve
problems at home as well as abroad. The astonishing successes in World War II
of what had previously been a fairly disorganized American scientific and technical community seemed to validate the notion of a special place for science,
and scientific thinking, in solving the problems of a democracy. This mantle
of scientific success extended to the social sciences, including the now supposedly objective findings of economics, sociology, and psychology. Flush with

new credibility, social scientists offered a rationale for the dramatic growth of government in fighting racism and socioeconomic inequality. In other cases, the connections between civilian science and the military-industrial complex were more direct, as when American city planners and managers turned to defense intellectuals for assistance in solving the intractable problems of the postwar city, from traffic and parking to poverty.

It would nevertheless be a mistake to fixate on military endeavors in understanding the role of science in the Cold War state. Starting with the legislation that established the AEC in 1946, American scientists fought to maintain a separate arena for civilian science. The existence of such civilian institutions as the AEC, the National Science Foundation (NSF), and the National Aeronautics and Space Administration (NASA) were critical to both policymakers' and the public's ability to see science as an objective, transparent way of knowing that could transcend differences between interest groups. This idea of "open science" sat uneasily next to the reality of a research infrastructure that was largely backed by state—particularly military—interests.

The space race is a classic example of these tensions. The rhetoric of the American approach to space exploration, particularly the Apollo missions that placed a man on the Moon in 1969, stressed civilian control, the international exchange of scientific ideas, and dramatic research results. At the same time, the Apollo program was designed to display and ensure American dominance; much of the underlying research was classified; and the majority of American space efforts were at least as much about producing missiles and spy satellites as they were about creating new scientific knowledge. The pretense of civilian leadership gave American image-makers the possibility of having it both ways.

These sorts of contradictions were not unique to the American case but, rather, followed inevitably from the position of science in the global postwar political environment. Consider, for example, the situation of physics in the Soviet Union. Official philosophical doctrine in the Soviet Union stressed the materiality of all things and the unity of theory and practice. The theoretical assumptions and culture of modern physics, with their emphasis on probabilities and uncertainty, proved difficult to reconcile with a materialist philosophy. For a time in the late 1940s it seemed as if Western-style theoretical physics might be prohibited in the Soviet Union. When push came to shove, though, Soviet authorities recognized that they could not develop a nuclear arsenal without the expertise and cooperation of the physics community. The desire to produce spectacular and timely results required the use of the best scientific information available, regardless of its philosophical pedigree, for any nation that wanted to compete on the Cold War stage.

The fundamental characteristic of Cold War science is the central role that the scientific enterprise came to play in the maintenance of the nation-state. Science and technology have, of course, always contributed to state power. In the Italian Renaissance, patrons requested that natural philosophers supply them with telescopes and astrolabes; two centuries later, the imperial governments of Spain, France, and Great Britain would send crews of naturalists to evaluate the commercial potential of the plants, animals, and minerals in their conquered lands. Nevertheless, this relationship underwent a fundamental change in the years immediately following World War II. For all their differences, leaders in both the Soviet Union and the United States agreed that massive displays of technological might were critical weapons in the international battle for hearts and minds. Scientific achievement had apparently won the war for the Allies; it would presumably be the critical factor in deciding the Cold War as well. This assumption transformed the scale and scope of scientific investigation and extended even to the ways that we talk about science.

The Cold War is now over, having been replaced with any number of emerging global conflicts. American science, particularly in the universities, is no longer dominated by military concerns; American policymakers no longer assign a preeminent role to science and technology in building international alliances. Nevertheless, the culture and practice of contemporary American science has been indelibly shaped by its phenomenal growth under Cold War auspices. The decision of many university administrators to sever campus ties to the military-industrial complex in the late 1960s and early 1970s had the unintended consequence of concentrating defense contracts in the hands of independent think tanks and for-profit corporations while simultaneously isolating scientists and military strategists from one another. In a peculiar irony, both the military and the universities have increasingly turned to corporate partnerships in hopes of raising revenue and driving innovation. This, too, is a legacy of the Cold War.

Students, instructors, and casual readers can approach this book in multiple ways. Chapters 1 and 6 offer snapshots of state-sponsored science in the early and late Cold War through the stories of two iconic projects: the atomic bomb and the Apollo program. Readers most interested in questions of how the Cold War affected the *practice* of science may be particularly interested in chapter 2, on the military-industrial complex, and chapter 3, on "big science."

In contrast, chapters 4 and 5 address the *uses* of science, particularly the social sciences, in foreign and domestic policy. Chapter 7 then turns to the fragmentation of the Cold War consensus that science, and scientists, had a duty to support state power. Chapter 8 considers how the role of science came to change in the 1980s as Communism began to crumble and economic fears increasingly took priority. Read in order, this structure will allow readers to encounter both the "can't miss" topics in the history of Cold War science—nuclear research, loyalty oaths, the growth of the military-industrial complex, the space race—and less obvious and familiar stories, such as psychological research on race and the campaign to eradicate malaria.

Although the chronologies of the chapters overlap, each moves the story slightly forward in time. In chapters 1 and 2, I focus on the early Cold War, the time during which federal support for science and technology was most strongly associated with military and defense needs. With the launch of the Soviet *Sputnik* in 1957, American policymakers' notion of scientific power grew to encompass prestige as well as weaponry. Chapters 3 through 6 explore how this broader notion of scientific accomplishment played out from approximately 1957 to 1969 in both the domestic and the international arenas. The inauguration of President John F. Kennedy in January 1961 marks a turning point within this second period, as the Cold War became an increasingly global race to capture the hearts and minds of the citizens of newly independent nations. The last two chapters trace the dissolution of the Cold War consensus in the late 1960s and the emergence of an alternative role for science in the 1970s and 1980s. These divisions are to a certain extent artificial, but they nevertheless serve to impose some useful order on what might otherwise be a sprawling account.

Every author must make decisions about what episodes to include and exclude from her account; this is particularly so in short, introductory volumes. As should already be clear, this is a work that focuses primarily on the American case. To maintain this focus, excursions beyond American borders are undertaken solely when they shed light on events closer to home. Similarly, although my definition of "science" is broad enough to include the social sciences and certain forms of technology, it does not encompass the medical sciences and therefore excludes a formal discussion of the expansion of the National Institutes of Health (NIH). Like academic science, academic biomedicine grew by leaps and bounds in the postwar period, also with generous support from the federal government. But if portions of the story are similar, changing attitudes toward individual rights and informed consent—in short, the human element—complicate the telling. Specialists from other fields may surely notice omissions from their own areas of expertise. Readers interested in learning more about the vibrant and growing field of the history of Cold War science are encouraged to begin their exploration with the books and articles suggested in the bibliographic essay at the end of this volume.

1 The Atomic Age

On August 6, 1945, the *Enola Gay* dropped the first atomic bomb—the Little Boy—over Hiroshima. The weapon, so new that it had never been tested, detonated according to plan 1,900 feet over Shima Hospital and devastated Hiroshima and its inhabitants. On August 9, a second atomic bomb exploded over Nagasaki. Though this bomb, the Fat Man, operated by means of a different mechanism, it was equally destructive, killing approximately 70,000 people. Faced with what seemed certain annihilation, the Japanese government surrendered on August 15, 1945. World War II had finally ended, but the Atomic Age had just begun.

Apart from the nuclear physics at its core, the atomic bomb and the devastation it caused did not immediately appear to be so different in kind from that caused by conventional weapons. The Allies' 1945 firebombing campaigns in Japan had already ravaged dozens of cities and left hundreds of thousands of civilians dead. The effects of radiation sickness were neither well understood nor expected at the time of the bombs' use. There was no question, however, that the atomic bomb was terrifying. The sheer scale of its destruction, along with its signature mushroom cloud, made the atomic bomb an unparalleled tool for psychological intimidation. The Japanese government's nearly immediate surrender after its use only increased the bomb's talismanic qualities. And—at least until 1949—it belonged only to the United States.

The atomic bomb was hardly the only high-tech weapon that won World War II for the Allies, but it has long played that role in historical memory. It was also critical to establishing the place of science in the Cold War state. The development of the atomic bomb required unprecedented cooperation between scientists, industry, and the military. Its creation was a massive project, bringing together thousands of scientists and technicians at multiple sites. The "secret" at its core suggested that both science and scientists held a special place in maintaining foreign policy supremacy. During the period of the United States' nuclear monopoly, the question of how to protect this so-called secret became a national obsession, with long-lasting consequences for the practice and culture of postwar science. Scientists, meanwhile, had begun to voice second thoughts about their role in developing the world's most destructive weapon. Who should control science, when scientific knowledge had itself become a weapon?

A New Kind of Weapon

The science of atomic weaponry is not particularly complicated. The splitting of an atomic nucleus—a process known as fission—releases large amounts of energy. Under certain conditions, the neutrons that are ejected during this process can cause other fission events in neighboring atoms, triggering a chain reaction. The tremendous amounts of energy released by a nuclear chain reaction can be used peaceably, as in nuclear power plants, or harnessed to create an explosion of terrifying force.

Scientists had speculated that fission might somehow be used as the basis of a weapon since shortly after its discovery in 1938, but the difficulties of securing enough fissionable material and creating a chain reaction seemed to make this a remote possibility, if it would be possible at all. Nevertheless, with the outbreak of hostilities in Europe in 1939, several countries—especially the United States, Great Britain, and Germany—started investigating potential military uses for atomic energy. In the United States, the effort was championed by German émigré scientists, including Albert Einstein, who were convinced that German scientists under Adolf Hitler were aggressively pursuing an atomic weapon.

At first, the American effort lagged. The military seemed more interested in atomic energy than in atomic weapons, and the committee charged with exploring the possibility of nuclear explosions accomplished little in its first eighteen months. In June 1942, faced with mounting evidence both that a bomb was possible and that the Germans were developing one, President Franklin D. Roosevelt approved a crash project to build an atomic weapon. The Manhattan Project, as it came to be known, would become the biggest collaborative scientific project the world had yet seen, requiring the efforts of more than 150,000 people in the United States, Great Britain, and Canada, and costing more than $2 billion.

The process of converting nuclear energy into a deliverable weapon required insights from the cutting edge of physics, chemistry, and engineering—and therefore the cooperation of the United States' most capable scientists and engineers. At the beginning of the war, uranium was the only known fissionable element. More than 99 percent of naturally occurring uranium is in the form of U-238, which is not particularly susceptible to a nuclear chain reaction. Its isotope U-235, on the other hand, makes an excellent starting point for a bomb. Therein lay the challenge: because U-235 and U-238 are chemically identical (the only difference between isotopes is in the number of neutrons, which has no effect on an atom's chemical behavior), separating them would require extraordinary means.

There was, however, another possibility. In 1941 Glenn Seaborg, a physical chemist at the University of California–Berkeley, discovered that while the breakdown of U-238 does not produce a chain reaction, it can produce plutonium, which *is* susceptible to a nuclear chain reaction. Although each individual reaction might not produce much plutonium, it would be much easier to separate plutonium from uranium than it would be to separate the uranium isotopes. The catch? For technical reasons, a bomb based on plutonium would require a much more sophisticated detonation system than one based on uranium. In short, the production of a weapon from either material would require solving difficult, if not impossible, scientific and engineering problems.

Since no one knew whether either the plutonium or the uranium weapon would work, the Army hedged its bets and developed both, creating a spectacularly massive federal research enterprise. For the uranium separation effort, the Manhattan Project would explore several different methods: gaseous diffusion, electromagnetic separation, and thermal diffusion. Each process required the construction of enormous production facilities and instant towns to house the scientific and technical personnel involved in the project. The industrial uranium separation activities were conducted at Oak Ridge, Tennessee, to take advantage of the nearly endless supply of cheap power produced by the nearby hydroelectric dams of the Tennessee Valley Authority (TVA)—powerful symbols of American technological might in their own right. The circular "racetracks" built there for the electromagnetic separation process used nearly 30 million pounds of silver. The gaseous barrier diffusion facilities required the construction of miles of pipes, valves, and tubes capable of withstanding the high pressure and corrosive effects of uranium hexafluoride gas. When the results from the gaseous diffusion process proved disappointing in 1944, workers built yet a third facility for an untried technique, this time for thermal diffusion.

On the other side of the country, scientists and engineers from the DuPont Corporation built a sprawling plutonium separation facility in Hanford, Washington. Meanwhile, scientists at the University of Chicago, Columbia University, the University of California, and many other academic campuses devoted their laboratories to working out the scientific problems at the center of the bomb. And at Los Alamos, New Mexico, the most famous of the

100-F Area at Hanford, 1944

The three reactors and chemical separation plants of the Hanford plutonium processing facility were distributed over half a million acres along the banks of the Columbia River in south-central Washington. The site had been selected both for its distance from major population centers and the availability of the Columbia River's waters; the water treatment plants used to decontaminate the hot uranium slugs could have supplied enough water for a city of a million people. At the height of construction in the summer of 1943, more than 40,000 construction workers lived on site, pouring concrete, laying brick, and building railroads to transport radioactive uranium rods. Besides plutonium, the separation facilities created enormous quantities of nuclear waste, stored in underground tanks. The entire process was so radioactive that the engineers could not enter the buildings; the railway cars were operated by remote control. Nearly seven decades later, the site is still contaminated.

■   Courtesy of the U.S. Department of Energy

Manhattan Project sites, scientists under the direction of theoretical physicist J. Robert Oppenheimer worked with military personnel to design deliverable weapons.

The clash of military and scientific cultures occasioned by the close quarters and high stakes of the Manhattan Project inevitably led to tensions. The entire project was cloaked in what was hoped to be an impenetrable veil of secrecy. Scientists working under pseudonyms refused to tell their wives what they were doing in the isolation of the desert. Security clearances and background checks were mandatory, and papers discussing nuclear fission disappeared from the scientific literature. At Los Alamos, Oppenheimer earned Army managers' wrath by insisting that the scientific personnel have weekly cross-departmental meetings to discuss their progress; the Army preferred a compartmentalized system in which knowledge was shared on a need-to-know basis.

As the war in Europe drew to a close, the question of who had the right to decide how this new scientific weapon should be used became a matter of contention. In the spring of 1945 a group of scientists at the University of Chicago under the leadership of Nobel Laureate James Franck argued that the bomb must not be used on civilians without a prior military demonstration. Perhaps because the university researchers had been focusing on theoretical questions of nuclear physics, the bomb's use as a weapon had, until now, seemed somewhat abstract. Among the project's scientific leadership at Los Alamos, however, the bomb's use in military operations was never really in doubt. The Franck Report was received politely by military leaders but ignored.

The first atomic bomb, a plutonium implosion device, was tested at Alamogordo, New Mexico, on July 16, 1945. The explosion released a blast equivalent to approximately 20,000 tons of TNT and sent a giant fireball into the desert sky, vaporizing the tower from which it had been dropped and turning sand into glass. Now, with a reliable detonation device and a steady stream of plutonium coming out of Hanford, the Manhattan Project could promise a steady, if limited, supply of bombs for late summer and early fall 1945. In the end, only two were needed—the Little Boy, a uranium device, and the Fat Man, a plutonium bomb.

Historians continue to debate exactly what factors went into the decision to drop the bomb. Given the growing tensions between the United States and the Soviet Union, some have argued that President Harry Truman and his advisors hoped to use the bomb to intimidate the Soviets. The successful Trinity test at Alamogordo had taken place at the beginning of the Potsdam Conference, where the Allied leaders had gathered to discuss their strategies for ending the war. Truman told Soviet general secretary Joseph Stalin only that the United States had developed a new and powerful weapon, not that it was an atomic device. Stalin asked no further questions—most likely, historians now believe, because he already knew about it. Other historians have pointed to the political imperative to use a weapon that had consumed unprecedented resources, or to the genuine belief among some leaders that its use might bring the war to a speedy close. And yet others question the premise of the need for a decision, asking whether the military would have seen the bomb's destructive powers as "special" given the extent of previous firebombing campaigns. Most likely the truth is some combination of all of these.

Regardless of the calculations behind the decision, the historical fact remains that the bombs were used, forever changing the role of science in war and foreign relations. Beyond the awesome spectacle of the bombs' destructive power, much about how they worked remained secret, but this much was clear: the United States, and only the United States, now had an extraordinarily powerful new weapon that drew its force from an obscure science. From the Americans' perspective, the question now became how to keep it that way.

Protecting the Bomb

The need to protect the secret of the bomb (or alternately, the need to protect the world from it) forced scientists and politicians to reconsider the rules of scientific openness. On the one hand, the three detonations at Alamogordo, Hiroshima, and Nagasaki had already revealed the answer to what many scientists saw as the most important question—whether it was in fact possible to use nuclear fission to fuel an explosive device. At the same time, plenty of secrets of the bomb's construction remained hidden from both the public and the broader scientific community. The technical expertise involved seemed to require a new, special, and ongoing role for scientists in developing and implementing weapons strategy, yet some scientists were beginning to question the propriety of dedicating their work to national interests rather than to the nation of science. Who had the right to determine the uses to which scientific knowledge could be put? Could scientific knowledge be contained within national borders? And how could the requirements of national security be reconciled with traditions of scientific openness?

Within a week of the bombs' use, the basic outlines of their scientific and technical details had been revealed to the public through the Smyth Report, named after its author Henry DeWolf Smyth. Major General Leslie Groves, the military head of the project, authorized the release of this information on the theory that, by doing so, the government could set the terms for what information was appropriate for public discussion. The report detailed significant information relating to nuclear fission, uranium isotope enrichment, and plutonium production. In the view of Groves and other military leaders the report omitted a good deal of crucial information, particularly in the realm of technological know-how. It pointedly did not discuss the plutonium implosion mechanism inside the Fat Man, nor did it acknowledge another key factor in the Americans' success—namely, that the United States and Great Britain, under Groves's personal direction, had secured the exclusive rights to more than 95 percent of the world's known uranium reserves. Many scientists who read the report felt that there was not much that remained secret after the report's publication, but most parties agreed that the United States could expect to maintain its nuclear monopoly for at least five years. Groves, who was one of the few people to know about the uranium stores, believed the potential period of monopoly to be much longer.

These different attitudes toward the existence of an atomic secret had significant consequences for both international and domestic politics. Many scientists, including Oppenheimer, believed that some sort of international agency—possibly even a world government—would be required to prevent atomic proliferation, in large part because so much of the science behind the project was either already available or would be obvious to qualified scientists. A number of schemes for the international control of atomic energy, ranging from inspections to full-scale disarmament, were proposed with input from scientists, but none succeeded, in part because those in charge of American foreign policy could envision few benefits from revealing the remaining secrets. For their part, Soviet officials saw little advantage to either international control or a research moratorium when the United States already had a weapon.

On the domestic side, a congressional bill introduced in October 1945 proposed placing nuclear research squarely in the hands of the military. Because its authors assumed that atomic weapons depended on the existence of an atomic secret, the May-Johnson Bill included strict security and secrecy measures. Oppenheimer and other leading scientific policymakers endorsed the bill, but it immediately encountered opposition from members of the broader scientific community, who had grown frustrated with wartime security measures. Those in the so-called atomic scientists' movement stressed civilian control, the free exchange of information, and international agreements. Their growing horror at the scale of the bomb's destruction, combined with their conviction that there was no atomic secret, left them convinced that an arms race would be inevitable without the oversight of some form of international agency. Besides lobbying Congress, the atomic scientists orchestrated a public information campaign on the virtues of civilian control of atomic energy. Their efforts were at least moderately successful, and by December it had become clear that the May-Johnson Bill could not pass.

The scientists' influence over atomic policy was to prove short lived. A new bill was drafted with input from the atomic scientists that seemed more sympathetic to their concerns. As originally written, the McMahon Bill stressed the peaceful uses of atomic energy, would have implemented civilian control, and acknowledged the possibility of future international restrictions on the development of military applications. Unfortunately for the scientists, the bill was introduced at a point of heightening tension between the United States and the Soviet Union. In February 1946, authorities announced the breakup of a Canadian spy ring that had passed atomic information to the Soviets. In light of this and other developments, the version of the McMahon Bill that eventually passed in the summer of 1946 bore more than a passing resemblance to the original May-Johnson Bill, especially in its emphasis on security and guarding atomic secrets. One central provision from the original bill did remain: the newly created Atomic Energy Commission would be staffed and run by civilians, not the military, and would support both pure and applied research in civilian as well as military applications.

Although the AEC's creation as a civilian agency was generally considered a victory for the scientific community, the agency would in practice focus primarily on matters of interest to the military. Since its management structure reflected these conflicting military and civilian goals, it might best be described as a quasi-military agency. With one exception, no scientists were appointed to its board; the AEC instead received scientists' input through the mechanism of a General Advisory Committee (GAC) that could advise but not implement decisions. David Lilienthal, the board chairman during the agency's first three years, had been a great champion of the international control of atomic energy but found his opinions frequently at odds with more conservative commissioners, Pentagon brass, and his congressional overseers on the Joint Committee on Atomic Energy.

With the establishment of the AEC in 1947, the Manhattan Project laboratories at Los Alamos, Oak Ridge, Berkeley, and Chicago (Argonne) were transformed into permanent national laboratories, complete with barbed wire and guarded gates. Scientists complained that continuing security restrictions and AEC ownership of fissionable materials were impeding the development of peaceful uses of atomic energy, such as power generation or medical uses for radioisotopes. By some estimates, three-quarters of research reports produced by the national laboratories during this time were classified. Nevertheless, the AEC did in fact sponsor many unclassified research projects at universities, particularly in biology, medicine, and high-energy physics. Emerging as it did from the fires of a new kind of weapon, nuclear research had an immediate and complicated relationship to scientific freedom.

Soon enough, it became clear that all of the elaborate security provisions had been for naught. From their expansion into Eastern Europe, the Soviets obtained uranium; from the Smyth Report and spy rings, they gathered information. On August 29, 1949, scientists in the Soviet Union detonated their first atomic bomb. The American nuclear monopoly was over.

A Calculus of Destruction

The revelation of the existence of a Soviet atomic bomb upped the ante for policymakers and nuclear researchers, ultimately forcing a clash between those who feared the military use of scientific knowledge and those who embraced it. In the years since the war, American research on atomic weapons had focused largely on improvements to existing fission weapons. Researchers at E. O. Lawrence's Berkeley Laboratory, for instance, continued to investigate ways to improve the separation of fissionable elements; those at Los Alamos worked primarily on improving the bomb's delivery mechanism. Following the Soviet atomic bomb explosion, however, a growing number of American defense advisors argued that these priorities were misplaced: a new kind of weapon was needed. The ensuing debate over the morality of developing the next generation of nuclear weapons split the scientific community and resulted in the creation of a second nuclear weapons laboratory. Moreover, the episode demonstrated the limits of scientists' power to influence the uses to which their research could be applied.

From the beginnings of the Manhattan Project, a small group of researchers had been investigating the possibility of developing a fusion bomb, otherwise known as a hydrogen bomb or thermonuclear weapon because of the intense heat needed to ignite the explosion at its core. Whereas fission splits atomic nuclei, fusion joins them using the same principles that produce helium from hydrogen in the core of the sun. The leading proponent of the hydrogen bomb, both during and after the war, was Edward Teller, a young Hungarian émigré turned U.S. citizen. Although Teller left Los Alamos for a position at the University of Chicago after the war, he continued to spend several months of each year at the weapons laboratory. The work on a hydrogen bomb had proceeded slowly for a number of reasons, not least of which were doubts about the theoretical possibility of constructing such a device. By mid-1949, however, Teller had become convinced that the bomb could be built. For Teller, an outspoken anti-Communist and technological enthusiast, the Soviet explosion signaled that the time was right for a crash program to develop a thermonuclear weapon.

The debate over the hydrogen bomb was not conducted in public; rather, the discussion occurred within the confines of the classified military and technocratic advisory channels that had come to dominate postwar science policy (see chapter 2). The AEC's initial response was to focus its efforts on a transition weapon known as a "booster" that combined a large fission event with a relatively small thermonuclear explosion. Such a device was thought quite likely to work but would increase the power of existing weapons only by perhaps ten. At Teller's urging, however, several other powerful figures—including members of the Joint Committee on Atomic Energy, military leaders, and Lewis Strauss, a hawkish AEC commissioner—began to press for a hydrogen bomb. Though a risky investment, a successful thermonuclear weapon offered the potential for an explosion at least a thousand times as powerful as contemporary atomic weapons. Since the question was in part deemed a scientific one, the AEC asked the GAC, headed by Oppenheimer, for a report.

Among the members of the GAC were some of the most influential members of the postwar science advising establishment. Besides Oppenheimer, the committee included James B. Conant, a leader in wartime research policy and president of Harvard University; Lee DuBridge, the director of the laboratory at the Massachusetts Institute of Technology that had developed radar and the proximity fuse; Enrico Fermi and I.I. Rabi, both Nobel laureates who had been involved in the Manhattan Project; Seaborg, the discover of plutonium; and Oliver Buckley, the president of Bell Laboratories, a major defense contractor. The Department of Defense, the Department of State, the White House, and Congress had repeatedly sought the advice of these men, individually and collectively, when making policy decisions. The scientists had significant experience with the realities of politics, and they believed—quite rightly—that their opinions carried weight.

The GAC report was an extraordinary departure from postwar assumptions about the role of science in supporting state power, all the more so given its members' political and military experience. Part 1 of the October 1949 report endorsed the booster program; part 2 declared that a crash program had a "better than even" chance of producing a thermonuclear weapon within five years. But in part 3 and subsequent addenda, the majority of the committee argued that the hydrogen bomb should not be developed, even if it were technically feasible, and possibly even if the Soviet Union decided to pursue one. A hydrogen bomb, they declared, could be used only to slaughter innocent civilians. It might be used as a weapon of genocide. Testing such a weapon would expose the entire world to dangerous levels of radioactive fallout. Developing it would tarnish the United States' image as a force for good and would only inspire the Soviets to pursue their own "super" bomb. In short, thermonuclear weapons offered few tactical advantages that could not be achieved through existing fission weapons and carried many strategic risks. As the addendum

UNIVAC Computing at Livermore National Laboratory

The development of the hydrogen bomb would not have been possible without the availability of high-powered, automated computers. Soon after Truman announced that the United States would pursue a thermonuclear weapon, the Atomic Energy Commission opened a second nuclear weapons laboratory at a defunct Navy air station in Livermore, California. The UNIVAC computer, pictured here, was one of the first pieces of equipment installed in the new Livermore National Laboratory facilities in January 1953. For the next twenty years, the laboratory (renamed the Lawrence Laboratory after the death of University of California–Berkeley physicist E. O. Lawrence in 1958) would host an intense exploration into the applications of nuclear explosives, from warheads and atomic artillery to civil engineering.

■ Courtesy of Lawrence Livermore National Laboratory

signed by Conant, Oppenheimer, DuBridge, and two others put it, "We believe a super bomb should never be produced. Mankind would be far better off not to have a demonstration of the feasibility of such a weapon."*

Their report was received with dismay. Many policymakers, including AEC commissioner Strauss, felt that the GAC had overstepped the bounds of its authority. Teller and Strauss lobbied strenuously for this position, arguing that loyal scientists had a duty both to investigate the secrets of nature and to

*Addendum to the General Advisory Committee to the U.S. Atomic Energy Commission Report of Oct. 30, 1949, as reproduced in Herbert York, *The Advisors: Oppenheimer, Teller, and the Superbomb* (San Francisco: W. H. Freeman, 1975), p. 157.

leave the decisions regarding the consequences of those findings to political leaders. Military officials were slightly more receptive to the GAC's report, fearing that a thermonuclear program would distract from the production of fission bombs. Ultimately, however, the Pentagon supported the hydrogen bomb, in part because military leaders saw little moral distinction between an atomic bomb that could kill hundreds of thousands of civilians and a hydrogen bomb that could kill millions.

By mid-January 1950, Truman and his advisors seemed to be leaning in the direction of a thermonuclear program. Once again, revelations of atomic espionage may have helped tip the scales: on January 27, Klaus Fuchs, a British scientist who had been privy to discussions at Los Alamos on the theoretical possibility of the hydrogen bomb, confessed that he was a Soviet spy, thereby confirming a suspicion of American and British intelligence agents that Los Alamos had a Soviet mole in its midst. On January 31, President Truman publicly announced that the United States would pursue the development of thermonuclear weapons. During the spring of that year, the Soviet presence in Eastern Europe plus Communist victories in China precipitated a shift in American foreign policy toward containment—the idea that Communist expansion could be stopped with military might. Truman's announcement, and the chilling of relations with the Soviet bloc and the Chinese, set the stage for the arms race that would come to dominate so much of Cold War science and politics.

While the American hydrogen bomb program was not without its theoretical and technical hiccups, it nevertheless proceeded quickly. In early 1951, Teller and mathematician Stanislaw Ulam conceived of a solution so "technically sweet," as Oppenheimer put it, that even he could no longer oppose the weapon's development. The creation of a second weapons laboratory in Livermore, California, accelerated the process. Within three years the Americans had developed a deliverable thermonuclear weapon. The so-called Bravo test on the Bikini Atoll in the South Pacific in 1954 yielded an explosion so massive (15 megatons) that it caused radiation sickness among a crew of Japanese fisherman working eighty miles from the test site. But as the dissenting GAC scientists had feared, the Soviets had lost no time developing their own hydrogen bomb. Seven months before Bravo, the Soviets tested a relatively small (400-kiloton) thermonuclear weapon; in November 1955, they detonated a larger (1.5-megaton) device much closer in design to the American bomb. This time, there would be no monopoly.

The discussions over how, and whether, thermonuclear weapons should be developed were to have important political consequences for scientists in both the United States and the Soviet Union. Conservatives in both Congress and the AEC had always been suspicious of Oppenheimer's political commitments; in 1953 and 1954 his doubts about the wisdom of developing a hydrogen bomb were used against him as evidence of his dubious loyalty. In 1954, the AEC—now under the chairmanship of Lewis Strauss—declared Oppenheimer a security risk and stripped him of his clearance. In the Soviet Union, too, some scientists expressed growing concern about the consequences of testing thermonuclear weapons, with similarly futile results. In 1954 Soviet physicist Igor Kurchatov circulated a letter to the Soviet leadership declaring the strategic folly of nuclear weapons and requesting a public debate about their merits. His advice was dismissed, but his request after the 1955 test to work solely on nuclear energy, rather than nuclear weapons, was granted. For the remainder of the 1950s and the 1960s, a dedicated group of international scientists vigorously campaigned to establish a nuclear test ban. Even so, it must be remembered that a much larger, if less vocal, portion of the scientific community continued to work on ever more lethal ways to design and deliver nuclear weapons. Some saw this work as a patriotic duty in the face of the Communist threat; some saw it as an opportunity to work on cutting-edge scientific problems; some saw it as just a job. Such weapons could not be built without scientists' cooperation, but scientists had little say in determining their use.

The atomic bomb held a central place in U.S. foreign policy and political culture in the years immediately following World War II—yet until 1950, the United States had barely a dozen bombs in its stockpile. Looking back at the extraordinary scientific, technical, and industrial achievements of the Manhattan Project, American policymakers found it difficult to believe that any other nation could compete with the wonders (and terrors) of American science. It is therefore perhaps not surprising that when American military and political leaders were faced with the startling news of the Soviets' own bomb in 1949, they once again assumed that high-tech scientific weaponry would offer better protection than diplomacy or international agreements. With Truman's announcement of the U.S. search for a thermonuclear weapon, the arms race had officially begun.

The atomic bomb was both a symbol of destruction and a symbol of the power of science. This power did not necessarily extend to the scientists themselves. Yet even as some scientists began to feel constrained by the requirements of working within the national security state, the institutional power of science flourished. Federal dollars from both military and civilian agencies poured into universities, industrial research facilities, and national laboratories; by 1950, the federal government was investing more than $1 billion annually in scientific research. Before the war, in contrast, annual federal expenditures on science were typically around $50 million, with much of the funding invested in agriculture and public health. Such a dramatic influx of money inevitably changed the course of research and the practice of science. We now turn, then, to Cold War science's most obvious legacy: the military-industrial complex.

2 The Military-Industrial Complex

At the end of World War II, many American leaders, civilian and military alike, credited the Allies' victory to the achievements of the scientific and technical community. These accomplishments had relied upon an unprecedented military mobilization of individual scientists, university resources, and industrial labs. The transition to peace, uneasy as it might be, raised the question for both scientists and politicians of whether it was possible to sustain this level of cooperation between scientific and military authorities outside the context of a national emergency. And even if it were possible, was such a diversion of industrial and intellectual resources desirable in a capitalist democracy?

The early years of the Cold War, from the end of World War II to the launch of *Sputnik* in 1957, witnessed a long and contentious public debate about the proper relationship between scientific research, the federal government, and the military. By the time the National Science Foundation (NSF), a public and civilian scientific research funding agency, was established in 1950, an alternative universe of defense-oriented funding had already taken shape. It is difficult to overstate the presence of military funding in university and industrial labs in the late 1940s and 1950s: by some estimates, more than three-quarters of all federal investment in scientific research came from a single military research agency, the Office of Naval Research (ONR), in the early postwar years. Not all of this money went toward weapons projects. A significant portion of ONR funds supported academic scientists pursuing so-called fundamental, or basic, research questions with little obvious military application. The funding for basic research at universities, however, was dwarfed by research and development (R&D) dollars spent on defense projects at universities, think tanks, and industrial and military labs. This emphasis on national security created new kinds of hybrid research institutions with their own elaborate rules of secrecy. By the mid-1950s, the wartime bonds between the military, industry, and the universities had coalesced into what would eventually become known as the military-industrial complex.

The Political Economy of Postwar Science

During World War II, American scientists, coordinated by the wartime Office of Scientific Research and Development (OSRD), had converted the abstract findings of biochemistry, chemistry, and physics into mass-produced antibiotics, radar, and the atomic bomb. But while military, political, and scientific leaders agreed that the nation needed the continued support of a vibrant scientific community, they differed in their assessment of how best to manage scientists and their work. Who should pay for science, and for what ends?

Military leaders such as General Groves believed that scientists proved most useful to the state when called upon to solve specific problems that required technical expertise—an attitude that became known as scientists "on tap, not on top." For scientists and engineers, including Vannevar Bush, the outgoing head of the OSRD, the lesson of the early history of the Manhattan Project was that military leaders rarely understood the potential of science to transform modern warfare. Scientists, Bush argued, must be free to pursue their research passions regardless of whether there were any obvious practical application, as their findings would inevitably yield technical innovation, economic stimulus, and medical breakthroughs. A third perspective, championed by populist senator Harley Kilgore, argued that the powers of science would best be harnessed under a national science program that could be held politically accountable to the needs of voters.

By the fall of 1945, the U.S. Congress was considering two contrasting proposals for a federal scientific research organization. The first, introduced by Senator Warren Magnuson, reflected Bush's vision of autonomous science. (It was drafted by one of Bush's aides.) Its proposal for a National Research Foundation excluded the social sciences, avoided discussions of intellectual property, and assumed that the agency would simply fund the most promising research, regardless of its congressional district. The bill sponsored by Kilgore, in contrast, envisioned a National Science Foundation that would coordinate federal research, sponsor educational fellowships, secure patents in the name of the government, and include a geographical formula for the distribution of research funds. Intended to support research in a wide variety of fields that might conceivably advance the nation's interests, Kilgore's proposal included funding for both the natural and the social sciences.

The most important distinction between the two bills, however, was in the attitude of each toward expertise. The director of Bush's foundation would be selected by a board of "qualified" (presumably scientific) experts, and proposals would be evaluated by scientific panels on the basis of intellectual merit rather than their contribution to society. In Kilgore's version, the director would be appointed by the president, and its board would include not only scientific authorities but also representatives from agriculture, industry, and labor. If Kilgore's proposal extended the hallmarks of New Deal centralized planning to postwar scientific research, Magnuson's bill was intended to erect a wall between scientists and the political process.

The chasm between these opposing views proved too large to cross in a single session of Congress. Nearly five years after its first introduction, the National Science Foundation Act of 1950 brought together elements of both perspectives. Both the NSF's director and its board would serve at the pleasure of the president, and the agency would be responsible for coordinating a national research plan. The agency would not fund research in the social sciences, but neither would it fund military projects. As signed into law, the NSF would focus on so-called pure or fundamental research projects, selected solely on the basis of intellectual merit without regard to geographical distribution.

The NSF would eventually become a powerhouse of scientific research funding. At the moment of its creation, however, it was a bit player. The original plans for the NSF included annual research budgets on the order of $100 million. By the time it was finally approved, however, the initial 1952 appropriation request had shrunk to $14 million, of which Congress funded only $3.5 million. In contrast, the U.S. Navy alone spent almost $600 million on R&D between 1946 and the start of the Korean War in 1950. While Congress debated the merits of a national research program, military agencies had continued to demand scientific research, and they got it. In the absence of a federal research agency, the AEC and the newly created Department of Defense (DOD) became the primary funding agencies for American scientific and technical research.

A unique institution in the Navy played a particularly important role in the postwar science funding landscape. Long before the war, the Navy had supported research at its in-house laboratories, especially the U.S. Naval Observatory and the Naval Research Laboratory. As of 1946, however, the Navy was largely shut out of what seemed to be the cutting edge of military science, namely, atomic research. Under General Groves, the Army closely guarded access to scientists and classified information. As some of these scientists returned to civilian life, the Navy saw an opportunity to form its own group of experts who might someday provide the winning edge over both foreign opponents and its interservice rivals. The result was the Office of Naval Research, authorized by Congress in 1946 in lieu of a National Science Foundation.

Although about half of the ONR's funds supported R&D work on mili-

Vannevar Bush and *Science: The Endless Frontier*

As World War II drew to a close, Vannevar Bush, director of the Office of Scientific Research and Development, began planning for the postwar organization of science. His resulting report, *Science: The Endless Frontier*, was released to President Truman in July 1945, less than a month before the first use of the atomic bomb. Starting with the example of penicillin, Bush argued that scientific progress is essential to both national security and the public welfare. As Bush put it, "Without scientific progress no amount of national achievement in other directions can insure our health, prosperity, and security as a nation in the modern world." This founding document of postwar science policy recommended the formation of a National Research Foundation headed by scientists, with divisions of medical research, natural sciences, national defense, scientific personnel, and publications. Although the eventual form of the National Science Foundation incorporated several of Bush's ideas, the organization's power and effectiveness in funding scientific research was initially limited by competition from defense-oriented agencies.

■ Courtesy of MIT Museum

tary objectives, the other half went toward basic, mostly unclassified research projects proposed and carried out by university researchers. For the Navy, the benefits of building relationships with scientists and universities outweighed the costs of paying for undirected research. University administrators approved of the ONR's generous overheads and willingness to pay for capital expenses; researchers enjoyed the lack of military oversight and minimal reporting requirements. Of the $86 million appropriated for the ONR's contract research program between 1946 and 1950, only about 10 percent went to projects directly related to naval applications. With such generous funding mechanisms and a willingness to support basic research, the ONR had become the de facto "Office of National Research" by the time of the NSF's eventual creation in 1950.

This situation was not without its critics. Detlev Bronk, the president of the National Academy of Sciences and a member of the Navy's Research Advisory Committee, questioned the propriety of the ONR's basic research program when the NSF was so clearly in need of political and institutional support. Similarly, Lee DuBridge, the president of the California Institute of Technology, warned in 1949 of the consequences that might result from allowing science to "exist merely from the crumbs that fall from the table of a weapons development program."* By 1953, however, having overseen the dramatic growth of Caltech with military funds, Dubridge had changed his tune. Caltech, he said, "would go broke very promptly" if basic research funds were stripped from the AEC and the DOD in favor of the NSF.† In 1954, President Eisenhower issued an executive order that conferred his blessing on this pluralistic approach to basic research: while the NSF should take over an "increasing" share of basic research, other agencies (that is, the AEC and the DOD) would continue to fund "mission-oriented" basic research.

While the above discussion of how best to fund basic academic research in the sciences is an accurate representation of the public debate over a major policy issue, it paints a somewhat distorted picture of the actual postwar research landscape. "Basic" research was only the tip of the funding iceberg. In 1950 alone, the DOD spent more than $500 million in R&D, of which less than 10 percent went to universities and nonprofits. Where the money did support universities, its effects were dramatic: historians have estimated that the AEC and the DOD supplied 96 percent of federal support for university research in the physical sciences in 1949. Nevertheless, the funds spent at universities were a mere fraction of DOD expenditures at defense contractors and in-house research facilities. In the 1950s, the DOD supplied, on average, a third of all R&D funds in private industry; in defense-oriented fields such as aerospace and electronics, nearly three out of every four research dollars originated with the military. Organizations such as Bell Telephone Labs, RCA, General Electric, Westinghouse, Raytheon, and Lockheed were, in a very real sense, extensions of the DOD. These trends were exacerbated by the outbreak of the Korean War in the summer of 1950. Within a single year, defense R&D doubled: in 1951, appropriations totaled $1.3 billion.

Given how controversial these military investments in civilian institutions would later become, it is important to note how little opposition they raised at

*Lee DuBridge as quoted in Paul Forman, "Behind Quantum Electronics: National Security as Basis for Physical Research in the United States, 1940–1960," *Historical Studies in the Physical and Biological Sciences* 18 (1987): 185.

†DuBridge as quoted in Daniel J. Kevles, "Cold War and Hot Physics: Science, Security, and the American State, 1945–1956," *Historical Studies in the Physical and Biological Sciences* 20 (1990): 259.

the time. The occasional individual researcher declined to accept military funds for personal or religious reasons, but these conscientious objectors—typically Quakers—stated their opposition in terms of personal morality rather than as a critique of the research system. And while some university administrators voiced concern about the long-term effects of accepting military contracts, most welcomed the financial support for university infrastructure. Of the estimated two-thirds of the nation's scientists and engineers who worked on defense-related problems in 1951, the majority felt confident in the importance of their work. Many had worn military colors in one of the two world wars and viewed their defense-related projects with patriotism and a sense of duty. Those less sanguine about the implications of a militarized society nevertheless believed that their weapons-related work might do more to prevent war than accelerate it. The logic of the arms race held that only the presence of ever larger and ever more destructive weapons could keep the peace. Scientists on the whole were no different from other 1950s-era Americans who could agree on the need to present a united front to a common enemy.

Hybrid Institutions

As the 1950s wore on, the lines dividing academic, military, and industrial research became more blurred. Campuses would increasingly house weapons development projects that could not conceivably be considered "basic research," while the AEC sponsored research in theoretical physics. By 1960, an outside observer would be hard-pressed to articulate the logic that housed basic science at AEC facilities, the development and operation of air defense systems at MIT, and theoretical economics at the RAND Corporation, a nonprofit think tank sponsored by the Air Force. A closer look at the work conducted at three of these hybrid institutions helps clarify how this close relationship between civilian and military science functioned within the broader security state.

The Applied Physics Laboratory and Project Bumblebee

Under the direction of the OSRD, wartime research had followed a standard division of labor in which university researchers handled concepts and design and industrial contractors managed development and production. There was one major exception to this rule: the Applied Physics Laboratory (APL) at Johns Hopkins University. The APL's scientists and engineers not only designed the proximity fuse, a radio-controlled device that exploded shells at a predetermined distance from their targets, but also oversaw its production and taught the Navy how to use it. As the war wound down, the Navy's Bureau of Ordnance took over the contract from the OSRD, and the APL became the center of the Navy's efforts to develop guided antiaircraft missiles under the name Project Bumblebee.

Administrators at Johns Hopkins had hesitations about nearly every aspect of the project, from its classified character to the unabashedly "applied" nature of the research. Nevertheless, the university found the initial $750,000 contract difficult to refuse. Soon the APL was operating within a new university division, the Institute for Cooperative Research, and—at the university president's insistence—with an industrial partner, the Kellex Company. University administrators hoped that the APL-Kellex relationship would be the first of many university-industrial partnerships in the postwar era. This was not to be. Not only did the APL fail to come to a working arrangement with Kellex, an experienced defense contractor, but other corporations had no interest in working with a laboratory that was jointly operated by a potential competitor. In 1948, the APL and Kellex parted ways, and the APL returned to the business of weapons development, this time as a separate division of the university.

As a Navy contractor, the APL thrived, contributing to the Polaris missile system, solid-rocket fuels, and satellites. Today, it employs nearly 5,000 people and operates a nearly $1 billion budget supplied primarily by the DOD, NASA, the Department of Homeland Security, and the National Security Agency. As an academic institution, however, it has been less successful. In the 1950s, unclassified research was quarantined to a "Research Center" that received approximately 5 percent of the overall budget. Researchers possessed neither tenure nor departmental affiliations, and the laboratory produced no graduate students.

Although its emphasis on weapons production set it somewhat apart from most other university-based defense contractors, the APL was in other ways typical of a new postwar scientific institution, the federal contract research laboratory. A federal sponsor—usually a defense agency—paid all costs and established programmatic goals. The university provided management services but otherwise had few expectations that the unit would perform in an academic context. In return, the university received administrative fees that could be used to support activities elsewhere in the university. Considering that the APL was located in Silver Spring, Maryland, more than twenty miles from the main campus, it stretches the truth to say that the laboratory was "at" Johns Hopkins: like many other federal contract research laboratories, the APL operated at a physical remove from its administrative sponsor. Similar arrangements prevailed at MIT, Stanford, Caltech, Cornell, and the University of Michigan, to name only a few. In some cases, as at the University of California, the budget of the associated laboratories came to dwarf that of the university.

High-Energy Physics at Brookhaven National Laboratory

Atomic research held a central and paradoxical place in the ideology of postwar science. The civilian status of the AEC was intended to signal that American scientists would explore the atom's peaceful purposes—atomic energy, radiation cures, new scientific insights—as well as its military uses. The realities of security restrictions and cost, however, made it unlikely that researchers working at universities could carry out the bulk of the AEC's work, even with generous contracts. The national laboratory system was thus developed as an alternative. Not truly "national," the six laboratories that would eventually bear the title performed more like regional hubs for scientists whose research required access to nuclear reactors or particle accelerators. While several of the labs, particularly Los Alamos and Livermore, effectively functioned as weapons facilities, others supported diverse research programs in physics, materials science, biology, and medicine. Brookhaven National Laboratory on Long Island became particularly known for supporting basic research that bore little obvious relationship to national security. While the majority of the national laboratories' research reports remained classified throughout the 1950s, most of the unclassified ones were released by scientists at Brookhaven.

Brookhaven was proposed and operated by Associated Universities, Inc., a consortium of several elite northeastern universities (Columbia, Cornell, Harvard, Johns Hopkins, MIT, Pennsylvania, Princeton, Rochester, and Yale) whose members felt shut out of postwar nuclear research in Berkeley, Los Alamos, and Chicago. Soon after opening in 1947, the lab had both a nuclear reactor and a particle accelerator, each of which was nominally open to visiting scholars. Designed to operate like a miniature university, Brookhaven was divided into academic departments of chemistry, physics, biology, and medicine and granted tenure to its senior staff—albeit tenure within the limits of five-year research contracts and security clearances. Even though Brookhaven's visiting scientist program fell short of initial expectations, the site's facilities were significantly more accessible to outsiders than those of the other national laboratories. During the summer of 1952, for example, almost a third of the scientific researchers onsite were visitors, including almost 100 graduate and undergraduate students. In 1955 the AEC declassified Brookhaven's research reactor, allowing visiting scientists, including foreign nationals, access without clearance.

Brookhaven's scientific leadership in high-energy particle accelerators illustrates how the AEC's justification for scientific research changed over time. The AEC's original interest in accelerators was a holdover from E. O. Lawrence's

Radiation Laboratory at Berkeley. Before, during, and after the Manhattan Project, Lawrence and his colleagues had used their cyclotron (a type of accelerator) to discover isotopes and explore the behavior of subatomic particles. Since most isotopes were not readily fissionable, and since most cyclotrons were small enough to allow universities to build their own, neither security nor cost automatically pointed to the AEC's involvement after the war. Nevertheless, the fact that General Groves had already agreed to support Lawrence's cyclotrons set a precedent: the AEC would invest in high-energy accelerators.

Brookhaven's first proposal for a 750-MeV (mega-electron volt) synchrocyclotron was rejected by the AEC in 1947 as not sufficiently large to justify its construction. The following year the agency approved Brookhaven's revised proposal for a 2.5-GeV (giga-electron volt) proton synchrotron. Over the course of its construction, the power of this device grew; it eventually came online in 1953 at 3.3 GeV. The Cosmotron, as it was known, allowed researchers to observe mesons, one of several subatomic particles theorized but rarely seen by 1950s-era high-energy physicists. But perhaps more importantly for the AEC's continued funding of particle accelerators, some defense strategists believed that a focused particle beam could be used to shoot down incoming nuclear weapons. Though many scientists dismissed the idea, it gained traction in the DOD in 1952 when researchers at Brookhaven discovered the principle of strong focusing, a technique that allowed accelerators to achieve much greater intensity at lower costs. When the resulting Alternating Gradient Synchrotron came online in 1960 at 33 GeV, it was the most powerful particle accelerator in the world. This device became a reliable subatomic particle generator, earning scientists credit for the discovery of the muon-neutrino, the J-particle, the omega-minus, and the charmed lambda particles—none of which have yet produced any military application.

SAGE and the System Development Corporation

Speed would have been of the essence in responding to a Soviet nuclear attack. As of the late 1940s, radar coverage of U.S. borders was spotty, and the data interpretation process was so slow that Soviet bombers might arrive before American interceptor planes could be engaged to shoot them down. The advent of intercontinental and intermediate-range ballistic missiles (ICBMs and IRBMs) in the mid-1950s only exacerbated the need for nearly instantaneous information. The main problem was that the computer systems of the day—both hardware and software—were in no way up to the task.

Defense analysts looked to Project Whirlwind, an experimental digital computer built by scientists and engineers at the Lincoln Laboratory, the Air

Force laboratory at MIT, as their best hope in breaking through this data log-jam. Though phenomenally expensive and beset with scheduling delays, Whirl-wind possessed the unique ability to process data in real time. In 1954, on the basis of a successful trial using a Whirlwind computer and radar data from Cape Cod, the Air Force authorized a full-scale computerized national air defense network, or SAGE (semi-automatic ground environment). As envisioned by the Air Force, SAGE "direction centers" would display aggregated, real-time radar data from twenty-three defense sectors. In the event of an attack, these centers would tell the Air Force where and when to send its interceptor planes.

The scale of SAGE is somewhat difficult to grasp. Only a decade earlier the Manhattan Project had dwarfed all previous scientific and engineering projects with its $2 billion budget; SAGE eventually cost $8 billion. Western Electric, RCA, General Electric, and Bell Labs all contributed to aspects of the project. IBM built the computers, each of which weighed 250 tons, but declined to develop the programming. Just assembling the hardware required the input of around 7,000 of IBM's employees—about 25 percent of its workforce—and the company simply could not handle the workload of programming it as well. When it became clear that no private contractor would be willing to take on this mammoth task, it was assigned to the RAND Corporation, the Air Force's nonprofit research institution based in Santa Monica, California. As of 1955, RAND estimated that it had about 10 percent of the nation's top programmers in its employ—all twenty-five of them. RAND therefore scoured the country for potential programmers, hiring approximately fifty college graduates each month. Hardly any of these young men had programming experience; instead, they were selected by their performance on a variety of psychological and mental aptitude tests. (Music teachers were believed to be excellent prospects.) IBM taught them how to use computers, and RAND taught them how to program.

In October 1956 RAND spun off its programming division into a separate nonprofit company, the System Development Corporation (SDC). By now, SDC had as many employees as the rest of the RAND staff combined. Three years later, SDC had more than 2,000 employees, including 700 programmers—about half of the country's supply. After the first SAGE center went operational in June 1958, additional installations opened at the rate of one every two months. As the programming work for SAGE wound down, SDC's employees fanned out into the market, working for private industry or other defense contractors. By 1963, SDC had more than 4,000 employees and 6,000 programming alumni. With an annual income of $57 million and almost fifty projects, it was by far the largest software company in the world.

SAGE transformed the nascent computer industry. The pioneering technologies developed to run it included video displays, parallel logic, analog-to-digital and digital-to-analog conversion techniques, real-time networking, and multiprocessing. The software built to power its computers included more than a million lines of code, representing a level of complexity unthinkable even five years earlier. Though many critics doubted whether SAGE would actually have worked in the event of a nuclear strike, at least two dozen real-time "command and control" systems were based on its direction center model in the 1950s and 1960s in the United States, NATO countries, and Japan. And in many cases, SDC—part defense-contractor, part programming university—programmed the software.

Secrecy and Security

The hybrid military-civilian nature of postwar science had dramatic implications for the production and sharing of knowledge. The DOD and the AEC were supporting scientific research because they believed that its findings could be useful in time of war; given the national security motivation, these findings would obviously need to be protected. The practices of secrecy begun during World War II—security clearances, classification codes, the compartmentalization of knowledge—became part of the culture of institutions operating with defense contracts. These practices affected not only the conduct of scientific research and collaboration but also the political process, as those scientists, congressmen, and members of the general public without a "need to know" were kept in the dark about weapons systems, spending priorities, and foreign policy.

Without security clearances, historians have only a limited understanding of the classified state. Even so, some outlines of the story have begun to emerge. Conservative estimates point to the classification of at least 7.5 billion pages of material in the fifty years following World War II; high-range estimates set the number closer to 2 trillion. A good portion of this classified material dealt with scientific information. Both the 1947 Atomic Energy Act and its successor, the 1954 Atomic Energy Act, designated all information related to nuclear weapons as "born secret," regardless of whether the person who created the information worked in a secure facility or possessed a clearance. Other areas of nuclear science, such as isotope separation, might not be classified but would become so if the information demonstrated the potential to be used in weapons production. Books, articles, reports, and even dissertations in physics, electronics, aeronautics, oceanography, and high-altitude physiology were classified on a regular basis, resulting in a "black literature" that could be accessed only in secure facilities by those with an appropriate clearance. Classified information might be something temporally specific and concrete, such as plans for troop movements, or it might be more abstract, such as the scientific explanation of the Teller-Ulam configuration for the hydrogen bomb.

The system of classification extended not only to scientific knowledge but also to knowledge about the practice of science. During the Manhattan Project, the University of California's contract with the Army for the administration of Los Alamos was classified "Secret," making it off-limits to senior administrative officials, including the university's contract attorney. Eventually three members of the Board of Regents received clearances so that the university could exercise at least a modicum of oversight over the lab it supposedly ran. Access to specific machines could require a clearance, as did, for example, access to all nuclear reactors until the mid-1950s. Even the reasoning behind government interest in scientific information could be classified. In 1950, for instance, the State Department released a report that stressed the importance of international scientific cooperation and collaboration in ensuring scientific freedom. This same report, however, also included a classified appendix that suggested that scientists' personal contacts might be a useful means to gather scientific intelligence at international meetings. For the program to be effective, the traveling scientists would need to be kept in the dark about the government's specific interests. Instead, other scientists—those with connections to the State Department and the Central Intelligence Agency—would debrief travelers upon their return and pass requested information along to the relevant agency.

This obsession with secrets and security complicated scientists' lives and imposed limits on their political activities. Beginning with the big science projects of World War II, American scientists found that obtaining and maintaining employment required security clearances. President Harry S. Truman's federal loyalty program, introduced as Executive Order 9835 in 1947, subjected all federal employees to security checks that extended to citizens' attitudes and associations as well as their activities. Security clearances were also required for university and industrial contractors who worked on classified projects, and in some cases for all employees of defense contractors, regardless of whether they actually worked on classified material. The security clearance process was vague, secretive, and idiosyncratic: junior applicants might spend six months waiting, only to be told that they had been denied, while high-profile research-

STATE OF MARYLAND - "LOYALTY PLEDGE"

CERTIFICATION OF EMPLOYEE OR OF APPLICANT FOR PUBLIC
EMPLOYMENT UNDER THE SUBVERSIVE ACTIVITIES ACT OF 1949

I, ..., do hereby certify that I am
not a subversive person as defined by law, namely, that I am not a person who is
engaged in one way or another in the attempt to overthrow the Government of the
United States, or the State of Maryland, or any political subdivision of either of them,
by force or violence, and that I am not knowingly a member of an organization engaged in such an attempt.

I further certify that I understand the aforegoing statement is made subject to the
penalties of perjury prescribed in Article 27, Section 536 of the Annotated Code of
Maryland.

In Witness of the truth of the statements hereinabove made I hereunto affix my

signature this .... day of .... 195..............

                                  Signature ....

Birth Date .... Home Address ....

                                                                     Sec. 131

State of Maryland Loyalty Oath Card

Anti-Communist sentiment reached a fever pitch at the dawn of the Cold War. Federal, state, and private employers frequently subjected applicants to security and background checks; some went so far as to require potential hires to swear a loyalty oath. This version of one such oath, required by the state of Maryland throughout the 1950s, specifically referred to an applicant's membership in a "subversive" organization—a murky term that could extend to labor and civil rights groups as well as Communist fronts.

■ From the Bentley Glass Papers; courtesy of the American Philosophical Society

ers might be able to slide through in a few weeks. Given that seemingly innocuous activities such as attending a labor rally or donating to a civil rights group could endanger a person's security status, scientists increasingly curtailed their political activities to ensure their future livelihoods.

In this feverish environment of secrecy and security, some scientists became targets of loyalty investigations. Both the U.S. Senate and the House of Representatives held hearings that sought to expose the role of Communists in government, science, entertainment, and education. Although the hearings yielded few convictions other than for perjury or contempt, they nevertheless caused enormous personal distress and public humiliation to many members of the scientific community, not least because many universities refused to hire or retain faculty who had been asked to testify. The receipt of a subpoena from the House Committee on Un-American Activities (popularly known as HUAC) also complicated the process of obtaining a security clearance.

The experience of Edward U. Condon, a prominent physicist well liked within the scientific community, illustrates how these mechanisms worked in practice. Condon had left the Manhattan Project because of disagreements with military leadership over security measures at Los Alamos. In 1945 Truman appointed him director of the National Bureau of Standards, a federal agency within the Department of Commerce that soon became an administrative node connecting military agencies and defense contractors. He first attracted the attention of HUAC in 1947, when the committee's chairman, J. Parnell Thomas, undertook an investigation into atomic espionage. Condon had supported civilian control of atomic energy and an internationalist view of science—both views opposed by Thomas. The committee soon released a special report declaring Condon to be one of the "weakest links" in atomic security, based largely on his pending security clearance from the AEC, his political views, and his involvement with alleged Communist front organizations.

The scientific community rallied to Condon's defense, arguing both that the claims were baseless and that Condon deserved an opportunity to defend himself. The Department of Commerce's loyalty board, meanwhile, cleared Condon, as did the AEC several months later. Even though Condon was never found guilty of any wrongdoing, the allegations would resurface for most of the next decade, following him into a career in the private sector as director of research for Corning Glass Works. In 1954, the Secretary of the Navy had Condon's security clearance revoked. Frustrated at once again having to demonstrate his loyalty and trustworthiness, Condon resigned, eventually finding a position as chair of the physics department at Washington University in St. Louis.

The security hearings for J. Robert Oppenheimer demonstrated a more direct link between clearance, access, and control. Oppenheimer was a central figure in the deeply interconnected world of postwar science advising, having served not only as the inaugural chair of the AEC's General Advisory Committee, but also on the DOD's Research and Development Board, the Navy's Research Advisory Committee, and the Office of Defense Mobilization's Scientific Advisory Committee, to name only a few of his high-level appointments. Though valued for his scientific brilliance and analytical mind, Oppenheimer had made a long list of enemies at the Pentagon, in the Congress, and at the AEC. In December 1953, President Eisenhower ordered that a "blank wall" be established between Oppenheimer and national security information. The following summer, over the vocal objections of many of the most prominent voices in the scientific community, the AEC declared Oppenheimer to be a security risk and formally stripped him of his clearance.

The scientific community reacted to these events with fear and disbelief. Even more shocking than the idea that Oppenheimer, the man who built the bomb, could be considered a security threat were the specific allegations of wrongdoing. Oppenheimer's detractors repeatedly pointed to the scientist's habit of offering advice beyond the realm of science. His moral opposition to the development of the hydrogen bomb was the prime example of his transgressions. In the face of a weak NSF and a virtually nonexistent civilian science advisory apparatus, the decision confirmed what many already feared to be the case: the only scientists who would be welcomed at the science advisory tables were those known to be in agreement with military goals. Since those who disagreed with investment in defense-oriented projects rarely had the clearance to discuss them, there was little room for dissent. So long as the state valued science primarily for its contributions to national security, this closed system would remain in place.

The 1950s-era Science Advisory Committee illustrates the limits of science advising in this context. Truman created this advisory board in 1950 as a counterweight to technical advice sponsored by the Pentagon, which often—perhaps not surprisingly—recommended the pursuit of ever more elaborate weapons systems. Although intended to provide an outside source of expertise, the membership of the committee largely overlapped with other defense-oriented science advisory councils, including among its members DuBridge, Oppenheimer, Bronk, Conant, and James Killian, president of MIT. Despite its high-profile membership, the committee was buried within the Office of Defense Mobilization, and Truman began ignoring it almost immediately after creating it. Instead, the committee's most lasting contribution was a report prepared after the Oppenheimer hearings on the United States' ability to withstand surprise attack. The 1955 Technological Capabilities Report, prepared by Killian, recommended crash programs in developing ICBMs and IRBMs and an increased investment in intelligence-gathering capabilities, including the pursuit of reconnaissance satellites. Not coincidentally, much of this research was already taking place at laboratories administered by Killian's own institution, MIT. Initially created to limit the growth of defense spending, the Science Advisory Committee ultimately encouraged the pursuit of new and expensive weapons programs.

Fueled primarily by federal dollars, American scientific and technical research underwent enormous expansion in the 1950s. Until the launch of *Sputnik* in 1957, however, only a tiny fraction of this investment originated with civilian agencies for nondefense purposes. But to what end? Did so much military investment change the path of American science, or did it just make it bigger?

The answer, of course, is both. Lee DuBridge had it right when he expressed his doubts over whether Congress would ever fund the NSF at levels comparable to military R&D. The only reason the government would provide such generous funding was if it believed that the research would provide some useful end. In theory, that useful end was national security (namely, weapons); in practice, the results were more varied. Biologists and physicians working on AEC grants explored the effects of radiation on the bodies of both soldiers and cancer patients. DOD and AEC dollars paid not only for bombs and rockets but also for discoveries of new subatomic particles. The software system that ran SAGE also provided the model for the first real-time computerized airline reservation system. The impossibility of untangling military and civilian uses of much of Cold War science and technology points to the larger militarization of American life under the perceived threat of Communist expansion.

A second consequence of the blurred boundary between military and civilian research was a growing anxiety over the status of "basic" research. The postwar scientific community defined "basic research" as that driven by disciplinary rather than practical concerns, while "applied research" sought to solve specific real-world problems. Research on the mechanisms of *Drosophila* genetics might be considered basic, while that on corn would generally be considered applied. The original proponents of the NSF, Vannevar Bush foremost among them, argued that both military and industrial patrons would neglect basic research in favor of projects with more obvious short-term gains. But Bush miscalculated the appeal of his claims to his intended audiences. Political leaders in Congress found it difficult to justify a blank check for theoretical research to their constituents. Some military and industrial leaders, on the other hand, accepted Bush's argument that the ultimate success of the Manhattan Project stemmed from basic research in the physical sciences decades earlier. The odd configuration of postwar science, in which a Navy agency provided most of the funds for fundamental research in universities and a civilian agency built bombs, was the result.

In this larger sense, Senator Kilgore's fears proved prescient. While military investment in basic research created a number of spin-off products that eventually drove economic growth—the computer and the transistor being perhaps the most obvious examples—such dividends were fortuitous rather than planned. As he predicted, federal dollars were concentrated at elite institutions on the two coasts rather than dispersed among the broader population. The massive investment in scientific R&D for defense purposes diverted funds from projects arguably of equal importance to the national interest—infrastructure, education, and public health, to name only a few. In the first decade of the Cold War, investment in science meant investment in national defense.

It would take the launch of a Soviet satellite to expand funding for science into areas that might bring international prestige as well as military dominance. Nevertheless, by the mid-1950s the scientific community was larger, wealthier, and better organized than ever before. American science had become big science.

3 Big Science

On October 4, 1957, the 184-pound Soviet satellite *Sputnik* traced a path of anxiety across the night sky. The launch was not particularly surprising to President Eisenhower and his small circle of military advisors, but it came as a shock to nearly everyone else. The problem was not so much the satellite itself as the rocket that sent it and its successor, the 1,120-pound *Sputnik II*, into orbit. A nation that could send up a satellite could, presumably, hurl much more dangerous items across time and space. Making matters worse, the United States' first attempt at launching a satellite two months later ended in smoke and flames. The Space Race had hardly begun, and the United States was already losing.

What had gone wrong with American science, the supposed cornerstone of the nation's security? *Sputnik*'s launch offered damning evidence for any number of arguments. Military leaders claimed that the embarrassing episode demonstrated the need for a Manhattan Project–style crash program for rocketry; advocates for the NSF saw the satellite as proof of the perils of ignoring basic research. Defenders of academic freedom pointed to the dangers of over-zealous classification schemes, while education reformers criticized the lack of rigor in high school science instruction. What everyone could agree on was the need for *more*: more funding, more scientists, and more coordination. And indeed, American investment in scientific research and training in both civilian and military institutions grew dramatically in the late 1950s and early 1960s, most visibly in the creation of NASA and the expansion of the NSF.

But rather than transform the character or quality of scientific research, this influx of dollars only accelerated changes that were already taking place. American laboratories, science classrooms, and disciplinary gatherings already looked much different in 1955 than they had in 1935. In 1961, Alvin Weinberg, the director of Oak Ridge National Laboratory, christened this new kind of enterprise "Big Science." The phrase seemed to capture perfectly any number of characteristics of Cold War science: the focus on large, expensive instruments; the corporate structure of scientific laboratories; the sheer number of scientists and engineers being produced; and of course the cost. The lone scientist working at his bench had given way to the research team collaborating on massive technological machines. And as the institutions of science grew, so did the need to staff and manage them. After a decade of uncoordinated growth, the launch of *Sputnik* provided the justification to finally establish something resembling a national science policy—albeit one that envisioned science, first and foremost, as necessary for national security and prestige.

A New Kind of Science?

The image of the scientist as a lone genius has always been at least a partial fiction. While natural philosophers like Isaac Newton could certainly theorize in splendid isolation, the history of science is equally rife with examples of large-scale, government-sponsored, coordinated projects, from natural history expeditions to efforts to map the stars. The establishment of industrial research laboratories at technology-based firms like General Electric and Kodak in the early twentieth century marked the emergence of another kind of institution, where teams of multidisciplinary researchers might work together on common problems. The need to establish collaborative research networks extended even to academic fields such as genetics, where researchers in the 1920s and 1930s found their own work increasingly tied up with that of others who studied similar species.

Even so, many scientific observers in the immediate postwar years felt that they were witnessing the emergence of something new, something different. Backed with federal funds, more and more university laboratories were popping up in the model of E. O. Lawrence's Radiation Laboratory at the University of California–Berkeley. When Lawrence's laboratory first took shape in the 1930s, its size, interdisciplinarity, budget, and operating structure were all somewhat unique. With a staff of approximately sixty people, the laboratory operated in shifts and maintained a budget ten times that of a typical university research laboratory. Most notable was the laboratory's intellectual focus. Instead of pursuing the questions that drove their disciplines, the physicists, chemists, physicians, and biologists who worked in Lawrence's lab centered their investigations on those questions that could be answered by the instruments at the lab's core. Having spent more than $1 million building a 184-inch cyclotron, Lawrence and his colleagues were inclined to use it.

In the interwar years, ambitious laboratories like Lawrence's assembled their funding from a variety of sources: philanthropic organizations, industrial sponsors, patent licensing, and even fee-for-service arrangements. These funds—having once seemed so generous—were dwarfed in the immediate postwar years by the free-flowing monies of the AEC, the ONR, and various other DOD agencies. Entrepreneurial researchers saw in this new funding structure an opportunity to continue the kinds of collaborative projects they had worked on during World War II, now with a compliant federal government willing to pay for the giant machines they had previously only dreamed of. The federal sponsors, too, realized potential benefits in growing and maintaining a cooperative pool of scientific experts, particularly after the start of the Korean War. Hence, the 1950s witnessed the proliferation of capital-intensive research facilities, including particle accelerators, nuclear reactors, digital computers, and radar systems (to name just a few).

For those scientists inclined toward teamwork, these instrument-based facilities offered exciting intellectual opportunities. The national laboratories, in particular, emphasized new relationships between physicists, engineers, chemists, and biologists, as cross-departmental groups came to rely on one another's expertise. At both Oak Ridge and Brookhaven National Laboratories, for example, laboratory directors established academic "departments," but each institution also featured "groups" focused on specific problems or projects. Not all of the interdisciplinary projects had their roots in physics; at Oak Ridge, one of the largest programs was Megamouse, an investigation of the long-term effects of radiation on mice that involved researchers from several different disciplines. Eventually, the interdisciplinary field of nuclear science, combining elements of nuclear chemistry, nuclear physics, nuclear medicine, and radiobiology, came to constitute its own discipline.

Nevertheless, some scientists found aspects of the new instrumentalism disconcerting. Centering a research program on a piece of equipment, whether a tool or a weapon, inevitably changed the rhythms of laboratory life and forced scientists to rethink the very question of what it meant to conduct an experiment. In no field were these changes more dramatic than in high-energy physics. By the mid-1950s, the staff of an accelerator-based laboratory typically numbered in the hundreds, with an accompanying crew of technicians, data processors, accountants, secretaries, janitors, security officers, and health monitors. It was no longer possible for physicists, working alone or in small groups, to build or operate the machines necessary to produce cutting-edge results.

Consider the experiences of Donald Glaser, the physicist who developed the bubble chamber particle detector. Glaser had completed his graduate thesis at Caltech using a relatively low-tech cloud chamber to investigate subatomic particles found in cosmic rays. This style of work appealed to Glaser, and his decision to accept a job offer at the University of Michigan was in large part determined by the university's willingness to allow him to establish his own lab rather than force him to join an existing synchrotron or cyclotron group. By 1953, he and a student had developed a countertop bubble chamber equipped with a high-speed camera that could document high-energy events. But capturing the short-lived and unpredictable nature of cosmic ray particles with this system proved challenging; instead, its sensitivity was ideal for coupling with accelerators, which emitted particles at predictable times.

Although Glaser continued to work with bubble chambers, he was quickly outpaced by Luis Alvarez's group at the Berkeley Radiation Laboratory. Alvarez preferred to fill his bubble chambers with liquid hydrogen instead of the diethyl ether used by Glaser, a decision that required the safety and technical expertise of engineers, physicists, and cryogenicists. Alvarez appointed a chief of staff to oversee the operation and used his old Manhattan Project contacts to ensure that assistance was forthcoming from other institutions. Whereas Glaser spent less than $5,000 building his tabletop prototype, Alvarez spent $1.25 million building a six-foot-long device. As predicted, Alvarez's machine was extraordinarily productive—so much so that it was soon producing many thousands more events than could possibly be analyzed by the existing staff. Alvarez solved this problem by adding a sophisticated and expensive automated system, the Spiral Reader, that could label, measure, and calculate the path of a bubble track. By 1967, with the help of the Spiral Reader and $13 million spent solely on data processing, the Radiation Laboratory was recording over a million high-energy events a year.

Both Glaser and Alvarez received Nobel Prizes for their work on bubble chambers: Glaser, in 1960, for the bubble chamber itself; Alvarez, in 1968, for the discoveries made possible by his modifications. There was no question that these machines had made new kinds of discoveries possible, but those scientists whose reputations were most closely tied to them nevertheless found the experience of using them somewhat trying. Glaser had joined the group at Berkeley in 1960; he soon left for molecular biology, a discipline he found more intellectually stimulating. Even Alvarez had become somewhat bored with the bubble chamber by the late 1960s, pointing out that technicians operated the machines, exposed the film, and analyzed the data. Which part of this, Alvarez wondered, constituted science?

Yet aside from these occasional expressions of frustration, most physicists welcomed the experimental and theoretical advances made possible by expensive accelerators and detectors. High-energy physics thrived within the institutional culture of the Cold War because the AEC—the agency that bankrolled it—believed in the inherent relevance of nuclear science to the national interest. What nuclear physicists wanted, nuclear physicists got. In the broader

Bubble Chamber

The availability of nearly unlimited federal funds for research in certain fields allowed researchers to construct large, expensive, capital-intensive instruments. The device shown here is a 72-inch liquid hydrogen bubble chamber housed at the Lawrence Berkeley National Laboratory at the University of California. Designed to detect subatomic particles, the bubble chamber became a virtual high-energy event factory when connected to a particle accelerator. Although instruments such as these made new scientific discoveries possible, the day-to-day tasks associated with their care and maintenance left many researchers feeling more like managers or mechanics than scientists.

■ Courtesy of Lawrence Berkeley National Laboratory

context of American science, however, such a close match between the interests of disciplinary leaders and their funders was rare.

A more typical example of how machines could constrain research programs to fit the agenda of those who footed the bill is MIT's Aeroelastic and Structures Laboratory. With Air Force and Navy funding, the ASL built a ninety-eight-foot-long "shock tube" designed to measure the effects of atomic explosions on aircraft aerodynamics. The shock tube served as the center of the lab's research and teaching efforts, and publications related to its results (mostly classified) formed the backbone of the new specialty of aeroelasticity. As military interests migrated to the aerodynamics of guided missiles, so too did the problems pursued by the researchers. This MIT lab, recognized as a world-class leader in aerodynamics with close ties to industry, notably did not explore problems of interest to civilian aviation, such as fuel efficiency or safety. These were, after all, not questions that could be asked of a shock tube.

In the social sciences, the machine was more often metaphorical. As in the physical sciences, defense-oriented agencies funded elaborate and expensive studies that came to shape the research methodologies of certain fields. One of the most influential approaches to 1950s-era communications research, for example, had its roots in an Air Force–funded study, dubbed Project Revere, that dropped millions of civil defense leaflets into remote American communities in Washington, Idaho, Montana, Utah, and Alabama. Social scientists at the University of Washington developed mathematical models to explain how these messages were received and spread among the broader population. Despite the fact that few civilian approaches to communication involve airborne propaganda, this stimulus-and-response model of communication came to occupy a central place in the postwar social sciences.

Researchers who work on machines ask questions that machines can answer. When the machines could answer questions of central importance to a discipline, as particle accelerators could for high-energy physics, machine-directed research seemed reasonable, even inevitable. In other fields, such as aeronautics and the social sciences, the diversion of intellectual resources from disciplinary questions to military problems was harder to ignore, even at the time. The move toward giant technological systems was concentrated in those fields—physics, aeronautics, electronics, oceanography, seismology, communications—that might plausibly have some utility to the federal government, the only organization large enough to foot the bill for capital-intensive scientific investigations. By the time that some critics began pointing to the intellectual hazards of instrumentalism as a cause of the United States' lagging performance in the space race, the Big Science approach had become so central to American scientific identity that it was unclear what the alternatives might be.

Manpower for Science

The massive technological projects embraced by military and quasi-military agencies required an enormous scientific and technological workforce to keep them running. To military and civilian leaders, the question of how to find enough people to perform these tasks—and more importantly, the right kind of people—took on crisis proportions as early as the Korean War. The launch of *Sputnik* drew renewed attention to the so-called scientific manpower emergency, with consequences both for science education and ideas about who might be a scientist.

By any possible marker, American universities experienced unprecedented growth and change in the decades following World War II. The amount of funding available, the number of students enrolled, the size of the faculty and administration, even the acreage covered by campus buildings—all increased dramatically in the 1950s and 1960s. As we have seen, much of this expansion was subsidized with defense dollars to ensure the government's steady access to both the products of science and the scientists who produced them. These university programs offered the additional benefits of familiarizing the next generation of scientists and engineers with military goals and technologies.

These structural changes in the American university produced particularly stark effects in graduate education. The idea that an academic career in science or engineering required an advanced degree was still relatively novel in mid-twentieth-century America, having been imported from the German research universities only at the end of the nineteenth century. As late as 1953, only two in five college teachers held a Ph.D.; a successful career in industry might still be possible with only four years of college education. Nevertheless, as American universities grew, so did the numbers of Ph.D.'s they produced. In the late 1930s, fewer than 100 research institutions granted about 3,000 Ph.D.'s. By the mid-1950s, the number of doctorates had tripled to nearly 9,000; by 1973, that number had again increased by almost a factor of four, with 35,000 students receiving Ph.D.'s. This extraordinary expansion was fueled in part by the baby boom—but in the case of science, demographics are only part of the story.

The funding priorities of Big Science actively encouraged the growth of graduate education. Both the AEC and the ONR explicitly considered the possibilities for training graduate students when distributing their funds. The particle accelerators that the AEC so generously supported, for example, were envisioned in large part as training facilities; recall that Brookhaven National Laboratory counted 100 students among its visitors in the summer of 1952. The ONR, the AEC, and, to a lesser extent, the NSF all offered fellowships to graduate students working in physics and other defense-related sciences. In the early 1950s, the AEC actually required that the recipients of its postdoctoral fellowships work on classified projects; three-quarters of the AEC's 1953 physics fellows went on to accept positions within the agency after completing their degrees. Major defense contractors, including Westinghouse, General Electric, DuPont, Raytheon, and IBM, similarly sponsored graduate fellowships in physics in hopes of cultivating enough workers to meet the requirements of their ever-growing military contracts.

The very scale of some postwar Big Science laboratories ensured the production of a steady stream of graduate students. Consider, for example, Charles Draper's Instrumentation Laboratory at MIT. Though technically a division of the MIT's department of aeronautical engineering, the I-Lab dwarfed its department with an annual budget of over $20 million and a staff of over 1,200 people by the early 1960s. This behemoth institution developed the inertial guidance systems used both in ballistic missiles and in Apollo spacecraft. But despite the lab's heavy R&D load, director Charles Draper envisioned it as a teaching laboratory. In the course of contributing to the lab's specialized weapons engineering systems, hundreds of military officers and civilians (but mostly military officers) completed theses on various aspects of fire control, radar, internal navigation, and instrument engineering. Indeed, the laboratory generated so many classified theses that it had its own thesis declassification officer for most of the 1960s. The scale of the I-Lab's operations made it somewhat unique, but its practice of assigning graduate students to analyze the reams of data generated by postwar instrumentation was increasingly common in the Cold War university.

Nevertheless, fears about a manpower shortage remained and were amplified by the surprise of *Sputnik*. Though they rarely offered a source for their claims, members of Congress declared that the United States was losing not only the Space Race, but also the race to produce scientists. President Eisenhower and his advisors agreed, announcing that investments in science education—at all levels—must exceed even investments in weapons and technology. In 1958, Congress passed an education bill, the National Defense Education Act, aimed at both high school and college science education. As its name made clear, this bill assumed a critical and unbroken link between science education, the number of scientists, and national defense. In its final form, the act included funding for merit scholarships, graduate fellowships, programs to improve science and math teaching in high schools (including funds to purchase laboratory equipment and supplies), and science curriculum reforms.

By 1962, university scientists had created curricular programs for high schools in mathematics, chemistry, physics, biology, and earth science. Most of these programs were modeled on the Physical Science Study Committee (PSSC), a physics program created by MIT physicist Jerrold Zacharias. Under Zacharias's leadership, the PSSC created instructional films, laboratory modules, and textbooks. At its peak in the early 1960s, the PSSC provided the materials for half of American physics courses. But while school boards readily adopted the program, it proved unpopular with teachers. The course was notoriously difficult for students as well as teachers, and the innovative materials failed to attract new students to physics courses. Most of the other curriculum reform programs generated similar complaints.

Educators' dissatisfaction with efforts to reform the scientific curriculum stemmed as much from a fundamental confusion about the nature of the "manpower crisis" as it did from university professors' unfamiliarity with the needs and capabilities of high school students. Were these programs, and other recruitment efforts, intended to attract the best and the brightest students into careers in science, to identify geniuses whose talents might otherwise be lost?

Or was the idea to recruit as many bodies as possible, to supply the machines of Big Science with technicians as well as managers? Was scientific and technical ability something that could be taught, or an innate skill to be identified among the lucky few? Employers, too, grappled with these questions as they attempted to identify promising candidates for scientific and technical positions. The computer programming industry, for example, turned to a combination of aptitude tests and personality profiles as tools for screening vast numbers of applicants, but employers found little correlation between high scorers and successful programmers. And despite continued claims of a personnel shortage, women and minorities with advanced training in science and engineering found that professional doors remained, for the most part, closed.

The nebulous goals of science curriculum reform and recruitment reflected a broader uncertainty as to what might make for successful science policy. Policymakers agreed that "science" was important to national security, but they had yet to articulate a coherent notion of what kinds of investments might generate the most useful returns. The consequences of this lack of knowledge became particularly clear in physics, the field that had most benefited from the influx of cash. Between 1861, when the first American Ph.D. in physics was granted, and 1940, American universities granted 2,400 doctorates in the field. By the late 1960s, they were issuing over 1,500 *a year*. By 1965, there were less than half as many jobs available as recent physics Ph.D.'s; three years later only one in four physics Ph.D.'s could expect to find a permanent position. In much the same way that the drumbeat of national security had produced enough nuclear weapons to destroy the world many times over, the constant rhetoric of scientific crisis had generated more physicists than the nation could possibly use. These facts pointed to both the exalted place of science in Cold War culture and the lack of consensus on how to manage the scientific enterprise.

Managing Science

With million-dollar budgets and thousands of employees, Big Science brought with it a host of management problems. During the mobilization of World War II, military laboratories usually fell under the management of the OSRD. The arrival of a nominal peace, however, brought with it an administrative vacuum. At individual research sites, scientific administrators such as Lawrence and Weinberg found themselves called upon to simultaneously fulfill the roles of accountant, publicist, and human resources manager. At the next level up the administrative chain, career bureaucrats learned to navigate the treacherous waters between scientists, politics, and the public purse. The reputations of such men as Vannevar Bush, the former director of the OSRD, and James

Killian, the president of MIT, were built as much on their administrative acumen as on their scientific achievements.

On the eve of *Sputnik* the United States lacked anything resembling a coordinated science policy. Each federal agency with a vested interest in science maintained its own advisory apparatus that advised administrators on issues related to its own specific interests. Had it not been for the fact that membership between the committees overlapped, this decentralized structure might have resulted in no information sharing between the branches of government whatsoever. Critics pointed to this decentralization as a prime cause of the *Sputnik* crisis. In 1957 the federal government was spending upwards of $5 billion annually on R&D—and what, besides ever more lethal nuclear weapons, did it have to show for it? The Army, Navy, and Air Force had each been pursuing their own rocket and missile programs at vast expense. When the first American satellite, *Explorer I*, finally breached the atmosphere in late January 1958, it rode atop an Army Jupiter rocket. The previous two attempts using the originally planned Navy Viking rockets had ended in humiliating failure on the launch pads. This raised the obvious question: Did the United States really need three rocket systems, or merely one that worked? Interservice rivalries had created a wasteful and duplicative system in which intradepartmental competition had displaced the overall goal of international scientific supremacy.

One might have expected the NSF to act as a coordinating agency, given that its founding charter granted its governing body, the National Science Board, the authority to coordinate scientific research not only for "national health, prosperity, and welfare," but also for national defense. But because the NSF was a new and underfunded organization, its board lacked the administrative structure and authority to impose its wishes on military organizations. Moreover, the NSF's first director, Alan Waterman, having previously served as the ONR's chief scientist, was not inclined to challenge military funding decisions. The result was that nearly a decade after the NSF's creation, no agency had attempted to prioritize federal investments in scientific research, military or otherwise.

In contrast, the year following *Sputnik* saw a flurry of attempts to create a federal science advisory apparatus. Congress created committees for aeronautics and space and seriously debated establishing a federal department of science and technology. The DOD created the Advanced Research Projects Agency (ARPA) to coordinate defense-related research, particularly on missiles and space, and gave it a civilian head, the director for defense research and engineering, who reported directly to the secretary of defense rather than to a specific branch of the military. The event that attracted by far the most fanfare, however, was President Eisenhower's decision in the fall of 1957 to assemble his own group of scientific advisors.

Eisenhower's appointment of James Killian as his special assistant for science and technology (a position commonly known as the president's science advisor) signaled a renewed recognition of the importance of scientific expertise in the federal government. In a city where proximity indicates influence, Killian was installed in a prime office in the Executive Office Building next to the White House. As part of his broad mandate to advise the president on all matters related to science and technology, Killian attended cabinet and National Security Council meetings and monitored scientific developments at other federal agencies, including the AEC, the DOD, the Central Intelligence Agency (CIA), and the Department of State. His views were augmented by those of an enlarged and reenergized version of the moribund Scientific Advisory Committee, now reorganized within the executive branch as the President's Science Advisory Committee (PSAC).

Beginning with its "Strengthening American Science" report in 1958, PSAC consistently recommended increased funding for basic research at American universities. In an echo of Vannevar Bush's refrain from *Science: The Endless Frontier*, PSAC contended that so-called basic, or undirected, research would prove the cornerstone of future economic success and national defense. A 1960 panel report prepared under Glenn Seaborg, a veteran of the Manhattan Project and then chancellor of the University of California–Berkeley, proved particularly influential in securing congressional support for basic research, graduate fellowships, and university infrastructure; the authors argued in no uncertain terms that academic research was of "absolutely critical importance to the national welfare." Among their recommendations was a plan to use federal support to double the number of "first-rate" centers of academic research over the next fifteen years.*

In the Seaborg Report and elsewhere, PSAC reiterated claims that leaders of the scientific community had been making since the end of World War II. What changed in 1958 was the willingness of policymakers to take their advice. Within a year of PSAC's creation, the NSF's budget appropriation had more than tripled, from $40 million to $130 million; by the first year of the Kennedy administration, it had doubled again. By 1964 NASA, a new agency (see chapter 6), was distributing over $100 million just to universities. The absolute level

---

*Both quotations are from the Seaborg report as cited in Roger L. Geiger, *Research and Relevant Knowledge: American Research Universities since World War II* (New York: Oxford University Press, 1993), p. 169.

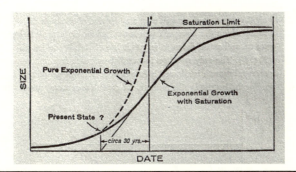

A Logistic Curve for Scientific Growth

In *Science Since Babylon* (1961) and *Big Science, Little Science* (1963), Yale historian Derek J. de Solla Price argued that American science was in a temporary period of exponential growth. Depending on whether the number of scientists, the number of journals, or the number of scientific articles were taken as data points, he determined that scientific knowledge was doubling every ten or fifteen years. As in nature, however, such exponential growth could not be expected to continue indefinitely. At some point, the growth curve would pass its point of inflection, after which growth would slow toward its natural ceiling. Price believed that the point of inflection for American science would arrive in another thirty to forty-five years. The concomitant crisis would lead to either a dramatic reorganization or the "death" of American science.

■ Derek J. de Solla Price, *Science Since Babylon* (New Haven: Yale University Press, 1961), p. 116.

of funding from the AEC and the DOD continued to increase, but the relative share of R&D funds provided by national security agencies declined. Much of this new funding was unrestricted, available for supporting graduate students or faculty, purchasing equipment, or even building laboratory facilities.

Under the leadership of Killian and his successors—George Kistiakowsky (1959–61) and Jerome Wiesner (1961–64)—American science entered a period of unprecedented support. But one might rightly ask, Were PSAC and the president's science advisors supplying advice, or were they advocating for a new interest group, Big Science? Is advocating growth the same thing as formulating science policy? By 1963, these questions had begun to worry Congress, which turned to the National Academy of Sciences (NAS) for input. Starting with astronomy, chemistry, physics, and the plant sciences, the NAS attempted to assess the past accomplishments, current trends, and future needs of various scientific fields. Inevitably, each of the resulting reports—prepared in each case by leaders of the discipline in question—made the case for increased support, ideally at a level of at least 15 percent annual growth. By this point, even some within the scientific community had begun to wonder whether a blank check truly represented the best way to foster scientific breakthroughs and technical innovation.

Industrial research managers faced similar questions of resource allocation and return on investment. In boardrooms, conferences, and specialized journals, research managers debated not only how much to invest in so-called basic research, but also how to supervise it. At the heart of their discussions were two different but closely related questions. First, could fundamental breakthroughs in scientific knowledge be planned? And second, what was the relationship of fundamental knowledge to industrial practice? At least through the 1950s, many corporate leaders preferred to err on the side of loose reins for scientists. As an illustration, consider the views of C. E. Kenneth Mees, the director of Eastman Kodak Research Laboratory, as expressed in a 1950 textbook on industrial research management. While Mees believed it was possible—indeed, desirable—to assign specific problems to a researcher, he thought it nearly impossible to expect the results to be produced on a schedule because "the scientist sets his own pace according to his enthusiasm and interest at the moment."* The president of Shell (Oil) Development Company agreed, writing in 1960 that "efficiency in research cannot yet be measured."† It was not unusual in the 1950s and early 1960s for industrial laboratories to allow scientific employees to spend 10 to 20 percent of their time on self-directed projects—with the assumption, of course, that any eventual marketable findings would accrue to the sponsoring company.

From one perspective, this reluctance to manage scientific employees could be seen as a major threat to shareholder profits. But once again, defense dollars changed the calculus of postwar research. With the DOD bankrolling a significant portion of the R&D budget at General Electric, Westinghouse, RCA, and other major industrial laboratories, industrial research managers had a cushion to absorb the occasional research project that didn't pan out.

Within the university, a new discipline—alternately known as "the science of science," "scientometrics," or, more simply, science policy—applied the techniques of the quantitative social sciences to make sense of science's growth. The most prominent advocate of this technique, Yale University historian Derek de Solla Price, used publication and citation analysis to argue that exponential growth was a normal and predictable feature of science. His 1961

---

*C. L. Kenneth Mees, as quoted in Steven Shapin, *The Scientific Life: A Moral History of a Late Modern Vocation* (Chicago: University of Chicago Press, 2008), p. 136.

†As quoted in Shapin, *Scientific Life*, p. 142.

book, *Science Since Babylon*, naturalized the growth of science, giving both science advocates and policymakers a way to talk about science policy without mentioning either the political context of the Cold War or the conflicts of interest inherent in the American science advisory system. In its assumption that the growth of scientific institutions and publications equaled the growth of scientific knowledge, Price's work was typical of a period in which managing science came to mean ensuring its expansion.

In short, American research administrators in the era of Big Science embraced the notion that scientific research required large federal investments that might or might not yield immediate, tangible returns. While these administrators might disagree on what proportion of research funds should be "undirected" or how long it might take for results to materialize, most agreed that at least some investment in basic research was necessary. There were, however, few agreed-upon yardsticks to indicate what might count as success. How does one calculate the return on an intellectual breakthrough? Ironically, the crisis of *Sputnik* forestalled the need for policymakers to answer this question: so long as the link between science and national security remained paramount, the federal dollars continued to pour in.

Scientific superiority remained a central pillar of American national security in the wake of *Sputnik*—but *Sputnik*-era analyses of science policy were remarkably devoid of discussions of scientific objectives beyond "growth" and "national preeminence." Scientific prestige now shared the spotlight with military leadership. Meanwhile, the rhetoric of American science—with its emphasis on disinterested civilian scientific leadership, unfettered growth, and the importance of basic research—belied an institutional structure funded in large part by defense and quasi-defense agencies. When American scientific leaders pointed to the "dangers" inherent in a scientific research policy skewed toward the programmatic goals of the DOD and the AEC, they were in part gesturing to the unintended similarities of this system to the Soviet planned economy.

From the American perspective, one of the most alarming aspects of Soviet scientific and technical success—whether in space or industrial production—was its potential to inspire the populations of unaligned nations. This battle over public sentiment became explicit in the early 1960s as Soviet leader Nikita Khrushchev pledged his support for the freedom of colonial peoples everywhere and U.S. president John F. Kennedy proclaimed that the path to peace lay in the "hearts and minds" of all peoples. American policymakers increasingly believed that winning the Cold War would take more than military hardware: it would require a wholesale demonstration of the superiority of the

American way of life. Scientific achievement became a central arena in which this battle for ideological supremacy played out. The United States might have lost the race to put the first satellite into space, but it would wage a fierce battle to produce the most Ph.D.'s, the most Nobel laureates, and the biggest particle accelerators.

Thus, while the relative proportion of federal R&D dollars tabbed specifically for defense dropped in the late 1950s and 1960s, it would be a mistake to assume that federal scientific investment moved away from Cold War concerns during the halcyon days of Big Science. The battle for hearts and minds turned *all* kinds of science—military and civilian, basic and applied, big and little— into proxy areas in which to demonstrate the superiority of the American way of life. The next two chapters turn to the effects of this ideological shift in both the wider world and on the home front.

4 Hearts and Minds and Markets

Science was instrumental in carrying out the geopolitical conflict between the United States and the Soviet Union, but the Cold War was driven by ideological differences. Leaders in both the United States and the Soviet Union believed not only that their own way of life was superior to that of their opponent, but also that only one system could ultimately prevail. The rhetoric and policy of the United States stressed individual rights and freedom within a global capitalist system. The Soviet Union subscribed to a theory of collective justice and economic socialism. That neither nation lived up to its rhetoric—and indeed, sometimes advanced policies that seemed diametrically opposed to its own ideals—should not distract from the fundamental fact that this was an ideological conflict.

The emergence of dozens of newly independent nations in the 1950s and 1960s destabilized this bipolar view of the world. Since the end of World War II, both the United States and the Soviet Union had been busily establishing and maintaining their separate spheres of influence—the United States in Latin America and Western Europe; the Soviet Union in Eastern Europe and, at least at first, in China. Within a decade, however, the global geopolitical landscape had changed dramatically, as one European power after another relinquished control over former colonial possessions. While some of the transitions were peaceful, others came after prolonged and sometimes violent resistance on behalf of the colonized. Between 1945 and 1960, approximately forty new nations came into existence in Africa, Asia, and the Middle East. But contrary to American and Soviet expectations, the leaders of many of the new nations seemed determined to steer courses separate from either capitalism or Communism. By the 1960s, it had become clear to American and Soviet leaders that neither words nor arms alone would be sufficient to win the allegiance of these new nations.

Science and technology played crucial roles in this battle for hearts and minds. In the United States, social scientists articulated elaborate models for economic growth as a means of kick-starting the development process in divergent cultures around the globe. They believed that scientific and technical expertise, along with democratic political institutions and education, offered the best route toward peace and prosperity for newly emerging nations. Globally, plans to bring American-style technological modernity to the Third World were implemented with varying degrees of success and failure, and with uneven levels of American support. But understanding why world leaders focused their efforts on "development" instead of other forms of assistance in the first place requires a discussion of the role that this concept played in the ideological conflict of the Cold War.

Ideology, Modernity, and the Creation of the Third World

As the conflict in what would eventually become known as the "Third World" began to play out, there was nearly universal agreement that development—a somewhat amorphous term that could refer to education, infrastructure, investment, industrialization, or expertise—would be essential for the nearly 800 million people in newly freed nations to reach the potential denied to them by colonialism. The question, of course, was what that potential might be. American foreign policy experts hoped that capitalist economic development, combined with liberal democratic institutions, would encourage stability and create strong ties to the American market. The Soviet Union believed that economic development and subsequent industrialization were necessary preconditions for the coming socialist revolution. Third World leaders—many of whom had spent time in the imperial cities of London, Brussels, Paris, and Lisbon—wanted to apply the best parts of technological modernity to their own nationalistic projects, particularly economic self-sufficiency from former colonial powers. Whatever their divergent goals and philosophies, each of these groups agreed that Third World nations required access to modern science and technology.

At least on the U.S. side, this interest in science and technology as an instrument of foreign policy was not entirely new in the 1960s. Private philanthropic organizations had funded international health and agricultural projects throughout most of the twentieth century. Following World War II, the United States had advanced several economic development plans for different regions of the world, each of which included programs for science and technology. The European Marshall Plan, for instance, not only provided over $13 billion in economic assistance for Western Europe, but also made possible the establishment of major European scientific institutions, including the European Organization for Nuclear Research (CERN). In South Korea, the Economic Cooperation Administration (a variation on the Marshall Plan for Asia) contributed significant resources to the rebuilding of the electric power network; a major point of conflict on the Korean peninsula both before and after the Korean War was the power supply, which had previously been located in the north. Similarly, President Truman's "Point Four" program—an assistance program for Asia, Africa, the Middle East, and Latin America—was intended, in Truman's words, to make "the benefits of our scientific advances and industrial progress available for the improvement and growth of under-developed areas."* Collectively, these investments (in science and otherwise) were meant to limit the appeal of Communism by raising standards of living and to deepen ties to American markets. Even so, U.S. investments in foreign science and technology in the early years of the Cold War were modest compared with domestic spending on scientific R&D.

Apart from its expansion into Eastern Europe, the Soviet Union offered surprisingly little assistance to developing nations in the early 1950s. Stalin's particular brand of Marxism held that nations could not transition to Communism without first experiencing industrialization. (Even imperial Russia was more "modern" than most emerging Third World states in the 1950s.) With a few exceptions, therefore, Stalin had discouraged international Communist parties that were operating within colonial states from fomenting revolution. Colonial rule, in his view, offered a more reliable route to modernity and industrialization—which would then be followed by the supposedly inevitable Communist revolution. Some of these international leaders chose to ignore Stalin's advice, eventually forcing the Soviet Union's hand.

Several events in the 1950s raised the level of attention Moscow paid to the Third World. First, Stalin's death in 1953 presented an opportunity for new Soviet leaders to advance different interpretations of Marxism. Nikita Khrushchev, in particular, believed that the Soviet Union had much to gain from partnerships with idealistic leaders in Africa and Asia, regardless of their commitment to classic Marxist doctrine. Second, the emergence of the People's Republic of China as an ambitious world power with an alternative model of socialist development forced Moscow to reckon with the possibility of ideological competition from the left. The Sino-Soviet Treaty of 1950 had, at first, seemed to assure a friendly relationship between the Soviet Union and the populous giant to its south. With the Great Leap Forward, however, Chinese Communist Party chairman Mao Zedong hoped to leapfrog China ahead of the Soviet Union by harnessing the power of the peasantry into a massive industrialized force. The Great Leap Forward turned out to be a humanitar-

*Inaugural Address of President Harry S. Truman, Jan. 20, 1949. Text and audio available online at the Harry S. Truman Library and Museum, www.trumanlibrary.org/educ/inaug.htm.

Prime Minister Nehru's Visit to the United States, 1949

The United States hosted Indian Prime Minister Jawaharlal Nehru on three separate visits between 1949 and 1961 in hopes of building a stronger alliance with the new nation. During the first trip, undertaken just two years after Indian independence, Nehru toured American farms, factories, and universities—sites that spoke to Nehru's commitment to achieving economic independence through technological modernity. The United States, in turn, publicized such events as a way of demonstrating American support for developing nations. Here Nehru (*center*) examines an ear of high-yield corn produced by farmers in Oswego, Illinois.

■ Photograph by Joseph O'Donnell; courtesy of the National Archives / U.S. Department of State

ian and administrative disaster, causing the deaths of an estimated 30 million people by diverting agricultural labor into ill-conceived industrial activity. At the time, however, the extent of this tragedy remained largely unknown to the rest of the world: what the world saw instead was a charismatic Mao who announced his solidarity with the peoples of Indonesia, Burma, Vietnam, Zambia, and Ghana. Meanwhile, the unanticipated success of homegrown Communist revolutions in Cuba and Vietnam placed the Soviet Union in the potentially awkward position of ignoring its own acolytes. By 1960, faced with competition from both the United States and China and increasing calls for help from far-flung allies, the Soviet Union had been pulled, ready or not, into Third World politics.

Of course, some Third World leaders recognized the advantages to remaining neutral in the Cold War contest. The very name "Third World" came about, in part, as a reference to a "third way" separate from capitalism or Communism.

In 1955, an impressive collection of African and Asian leaders representing a combined 1.5 billion people convened a meeting of so-called nonaligned nations in Bandung, Indonesia. There, India's Jawaharlal Nehru, Indonesia's Sukarno, and Egypt's Gamal Abdel Nasser pledged Third World solidarity as a means to overcome poverty, racial discrimination, and threats to national sovereignty. Six years later, India, Indonesia, Egypt, Yugoslavia, Ghana, and Algeria formalized their partnership as members of the Non-Aligned Movement. But while the attendees at Bandung and subsequent conferences of nonaligned nations gave much attention to economic cooperation, few concrete economic proposals resulted, in large part due to the commonalities of Third World needs and resources. These were nations rich in key natural resources—uranium, oil, precious metals—and poor in technologically sophisticated industrial and consumer goods. The one notable exception, the Organization of Petroleum Exporting Countries (OPEC), created as a partnership of oil-rich nations in 1960, serves as a reminder of how much effort these states invested in leveraging their precious resources within a global Cold War economy.

An important political consequence of Bandung was the increasing ability of resourceful Third World leaders to make the United States and the Soviet Union compete for their favors. In 1956, for example, Nasser announced his plans to nationalize the Suez Canal, a critical link between the Mediterranean, Africa, and South Asia that had previously been controlled by Great Britain. Nasser's decision was spurred, in part, by the United States' withdrawal of support for the construction of the Aswan High Dam, a hydroelectric project modeled on the TVA, after Nasser had purchased weapons from Czechoslovakia and offered diplomatic recognition to Mao's China. When the Soviet Union agreed to put up funds for the dam, Nasser nationalized the canal. What followed was, from the American perspective, a diplomatic disaster. Israel, France, and Great Britain invaded Egypt; the United States threatened the invaders—its own allies—with economic sanctions; and the Soviet Union rattled its nuclear arsenal. In the end, Nasser kept his canal, and the United States was forced to reckon with increased Third World negotiating power.

American, Soviet, Chinese, and Third World leaders therefore agreed: the global political future depended on development. As Nasser's experience with the Aswan High Dam demonstrated, Third World leaders needed international support to supply the funds and, in some cases, the technical expertise to build the infrastructure projects they associated with economic independence and national glory. The fierce competition between the superpowers meant that a clever leader could use the threat of shifting allegiance to extract needed goods and services from Washington, Moscow, Beijing, or some combination of the three. At the same time, the mostly failed attempts of Third World countries to establish trading partnerships among themselves after the Bandung conference served as a reminder of the products and services that these new nations lacked. Although differing in their details, both the Soviet and Chinese economic examples suggested models for the rapid industrial development that many Third World leaders desired. It was left to the Americans to develop their own theory of economic development, one that promised sustained economic growth within a liberal democratic system.

Economists and the Developing World

American federal investment in the social sciences never matched the sums poured into the physical sciences and engineering, but certain policy-oriented areas of sociology, political science, and economics thrived during the Cold War. Especially in demand were scholars who promoted a quantitative and supposedly value-free approach to understanding developing societies. President John F. Kennedy relied heavily upon the advice of social scientists at the vanguard of development theory in creating policies aimed at winning the friendship of newly emerging nations in Asia, Africa, Latin America, and the Middle East. By encouraging the development of foreign investment, scientific infrastructure, and liberal institutions, American social scientists hoped to guide these nations into a peaceful, prosperous, and, above all, modern future.

American modernization theory derived from postwar attempts in sociology to construct a comprehensive theory of society. The most influential of these theories, developed by Talcott Parsons and Edward Shils, held that stable societies depended on a close fit between social structures and culturally specific patterns of action. A society at equilibrium featured social, political, and economic institutions that matched the cultural values of its members. Parsons and Shils proposed that most important human attitudes and actions could be split into dichotomies, called "pattern variables," that defined how individuals understood their world. For example, an actor might be oriented toward the good of the self, or he might be oriented toward the good of the collective. Since individuals and societies tended to fall into easily recognizable clusters, pattern variable analysis could be used to locate any given society on a linear scale from the "traditional" to the "modern." Armed with questionnaires and checklists, researchers inspired by this work fanned out across the globe, assessing levels of literacy, urbanization, political participation, exposure to mass media, and use of technology as an index to a given society's readiness for modernity. An intervention in any one social area, if conducted carefully so as not to upset the social equilibrium, would inevitably lead to changes in others, allowing the possibility—in theory at least—of a predictive science of development.

By the mid-1950s, this schematic understanding of society had become popular among a disciplinarily diverse set of social scientists, particularly those with close relationships to federal agencies. The State Department and the CIA, for instance, found this model useful for comparing and forecasting the political future of nations with radically different social, political, and cultural institutions. Funding from these agencies, along with support from private philanthropic organizations, allowed the creation of such interdisciplinary research centers as MIT's Center for International Studies (CENIS), where psychologists, political scientists, economists, and communications specialists brought their expertise to bear on problems of foreign policy. In response to clear demands from their sponsors, these centers produced practical recommendations on the most pressing issue for developing nations: managing the "transition" between traditional and modern societies.

The crux of the problem for American policymakers and theorists was that the advent of modernity, in and of itself, did not guarantee the formation of a liberal democratic system. Pattern variable analysis suggested that both the United States and the Soviet Union represented forms of modernity. Moreover, the recent experiences of Nazi Germany were interpreted as evidence that rapid industrialization in a society not yet socially or psychologically ready for it could yield devastating political consequences. The American theorists, whose own ideological biases went mostly unacknowledged, assumed that their own version of modernity was "normal," while the Communist version was "pathological."

A classic example of this attitude can be seen in the interpretation by CENIS political scientist Lucian Pye of the appeal of Communism in postwar Malaya. On the basis of his interviews with disillusioned former members of the Malayan Communist Party, Pye concluded that Communism appealed to people who felt adrift from traditional values but were unable to attain success in the modern system. Their membership in the party, while giving their lives order and meaning, was ultimately a form of psychopathology. By using such seemingly scientific methods as survey instruments, pattern variable matrices, and quantitative analysis, social scientists surrounded their conclusions with an air of inevitability. This scientific ethos made it possible for American developmental theorists to draw a contrast between their own supposedly objective foreign policy interventions and the more ideologically driven actions of their Communist rivals.

Perhaps the most politically influential of the development theorists was

Rostow's Theory of Economic Take-Off

Walt Whitman Rostow—the most politically influential of the 1950s- and 1960s-era development theorists—believed that sustained economic growth was the key to modernization. Using historical economic data from the United States, the United Kingdom, Russia, Germany, and Japan, Rostow argued that a sustained, intense period of 5 to 10 percent net investment in an economy would inevitably lead to a period of "normal" economic growth. Once a nation reached this so-called take-off point, its economy would expand in predictable ways for approximately sixty years before reaching maturity. Post-maturity, a capitalist society would enter a period of "high mass consumption" in which excess funds were invested according to personal choice. Rostow characterized the Soviet alternative to high mass consumption, in which the government dictated economic choices, as pathological.

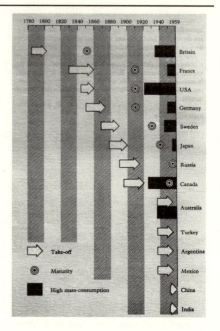

■ W. W. Rostow, *The Stages of Economic Growth: A Non-Communist Manifesto* (New York: Cambridge University Press), p. xii.

Walt W. Rostow, an economist at CENIS. Rostow and his colleagues had prepared background documents for Democrats in Congress throughout the 1950s. With the election of President John F. Kennedy in 1960, Democrats controlled the White House, and Rostow would go on to hold a number of senior foreign policy positions in the Kennedy and Johnson administrations, ultimately serving as Johnson's national security advisor from 1966 to 1969. His views did much to shape the mission and structure of the many 1960s-era international development projects that collectively increased U.S. spending on foreign aid by 25 percent during the Kennedy administration. Ambitious, idealistic, high-profile agencies such as the U.S. Agency for International Development (USAID) and the Peace Corps attempted to put into practice the belief of modernization theorists that all nations, everywhere, given sufficient financial and technical assistance, could be shepherded into a future that looked not unlike America's present.

These ideas are spelled out most clearly in Rostow's *Stages of Economic Growth: A Non-Communist Manifesto* (1960). Like most economists of his day, Rostow assumed that the key to modernization was economic growth. He proposed that the key moment in a society's economic development was the "take-off," the point at which economic growth becomes a "normal condition." Prior to the take-off, economic growth was limited by lack of technology, education, or cultural norms; post take-off, economic growth would be limited primarily by the choices a society made about how to spend its excess capital.

The key to achieving take-off, according to Rostow, was increasing net economic investment from approximately 5 to 10 percent. The problem for "traditional" societies was that they were limited in their ability to increase production and therefore to free up additional sums for investment. An agricultural society, for example, might be able to expand the number of acres planted, but it could do little to increase the yield on the existing acreage without such modern innovations as hybrid corn or artificial fertilizer. Rostow described this problem as a "pre-Newtonian" mindset toward science and technology.

But modern technology alone would not be sufficient to ensure take-off. A true take-off would require a rejection of rural attitudes and agricultural practices in addition to the establishment of new economic institutions and, usually, a powerful centralized state. Using historical evidence from Japan, Germany, and Russia, Rostow argued that the underlying social changes necessary for a successful economic take-off were usually driven by external threats, particularly military threats from more advanced nations. And therein lay the challenge and opportunity American foreign policy planners perceived in the Third World. The advent of nationhood for the former colonial states brought with it an unparalleled opportunity to yoke nationalism with investment in the public good, but it also eliminated much of the foreign expertise and investment that would be needed (according to this interpretation) to drive development. What would take the place of colonial dollars and expertise?

To liberal American foreign policy experts, the answer seemed obvious: American dollars and expertise. For example, in Latin America—an area many policymakers believed to be on the threshold of economic take-off—the Alliance for Progress Program not only pledged more than $20 billion in private and public investment over the course of a decade, but also flew Salvadorean businessmen to small towns in Missouri where they could witness firsthand the entrepreneurship and initiative of American small business owners.

Peace Corps Volunteers in Chad

President John F. Kennedy's creation of the
Peace Corps in March 1961 symbolized the
optimism of U.S. development theorists
for the future of the Third World. Thousands of idealistic young Americans fanned
out across the developing world in hopes
of improving America's image and making
the world a better place. Typical projects
focused on building political and technical capacity as a means to better prepare
transitional societies for economic development. Although optimism (both for the
program and on behalf of its volunteers)
soon faded, the Peace Corps remained a
central tool of American diplomatic relations with the Third World throughout
the 1960s. The volunteers depicted here are
working with a team that is drilling a water

well in Chad, one of the world's poorest
countries, in 1968.

■ Courtesy of the Peace Corps

The Peace Corps program similarly focused on contacts between individual
Americans and local communities. When American volunteers taught science
classes in Ghana or dug wells in Chad, they were sharing their "can-do" mindset along with their scientific and technical knowledge. Collectively, these programs transferred hundreds of billions of dollars to developing countries in
the 1960s and 1970s; the programs' administrators envisioned the contacts
between Third World elites and Western experts to be as critical as the actual
transfer of funds.

Yet, even in Rostow's theory of development, there remained a risk. According to his analysis, a "mature" economy could be either Communist or
capitalist. From the 1920s on, Americans had spent heavily on personal consumption, while the Soviets shunted all available funds into heavy industry. If
economic growth itself was the goal, the American system, with its emphasis
on personal financial choice, would be at a disadvantage almost by definition.
Left to their own devices, Soviet citizens, too, might invest their extra rubles
in washing machines, televisions, and automobiles instead of steel plants and
military technologies. What then, was the economic appeal of consumer capitalism for developing nations? Rostow's answer was to cite the inherent moral

superiority of a democratic system in enabling individuals to achieve their full "human potential." "An economy," he wrote, "is an instrument for a larger purpose."* The job of development experts was to provide the financial and technical resources that would allow the citizens of all societies, in both developing and developed nations, to live lives full of dignity and balance and free from want.

Rostow's explanations, like those of most development experts of his generation, left many questions unanswered about how, exactly, science, technology, and economic growth would promote a democratic system. Too often, the questions went not only unanswered, but also unasked. And even at the time, some critics argued that U.S.-backed development plans, focused as they were on establishing economies powered by mass consumption, were aimed more at expanding American markets than at solving Third World problems. Nevertheless, these were the idealistic attitudes upon which countless "Development Decade" projects were built. Understanding how these theories played out in practice, for better and for worse, requires a closer look at representative projects.

Development in Action: Three Snapshots

Western development projects took many different forms, in locations as varied as African cities, Middle Eastern deserts, and Southeast Asian rice paddies. Whatever their specific goals, the vast majority of development projects started with the assumption that Third World nations were in need of Western methods and attitudes, and that both would be readily transplantable to foreign soils. In practice, this was rarely the case. As the success or failure of an individual project almost always depended on local circumstances, what follows can only begin to suggest the complexity of factors that shaped efforts to transition Third World societies into modern capitalist nation-states. Even so, it is fair to say that projects based on science and technology tended to focus either on building infrastructure, such as roads, bridges, and dams, or on building human capital, through health, education, or expertise. Frequently, as in each of the examples below, the projects became a screen on which each participating institution could project its own goals and desires.

Building Dams In Tanzania

Technological enthusiasts looked to hydroelectric dams as agents of modernity for most of the twentieth century. Both the Soviet Union and the United States

*W. W. Rostow, *The Stages of Economic Growth: A Non-Communist Manifesto* (New York: Cambridge University Press, 1960), p. 103.

touted their success with hydroelectric power to audiences across the globe, but it was the American example of the Tennessee Valley Authority that held particular sway in the imaginations of ambitious Third World leaders. From the 1940s through the early 1970s, American image-makers used stories of the TVA's transformation of the Tennessee River valley as a powerful example of the possibility of state-driven social transformation within a capitalist system. No diplomatic tour of the United States was complete without a visit to a TVA dam—as one State Department representative admitted in 1949, "it's the first thing we think of."* The enthusiasm of both local leaders and international experts for such projects led the United Nations in 1956 to identify river development as a central component of economic growth for developing nations. The boom in dam building over the next fifteen years drew heavily on these visions of American development even when the projects operated without American funds.

Throughout the Third World, advocates for dams hoped that harnessing river power would make new lands available for agriculture, supply water for drinking and irrigation, and provide power for vast regions of the countryside, as it had in the American South. But as had been true in the United States, these utopian technological visions glossed over the potential pitfalls of converting water to power. Even in a best-case scenario, building a successful dam requires cooperative weather, the relocation of towns and villages, significant engineering prowess, and some semblance of an electrical grid. Large dam projects forever alter the local landscape, displace residents, and change the rhythm of local life. Sometimes these challenges prove so daunting that the dam is never built at all. Such was the case with Stiegler's Gorge in Tanzania.

Tanzania is an East African nation on the shores of the Indian Ocean, formed in 1964 from the unification of the former British and German colonies of Tanganyika and Zanzibar. The Rufiji River basin forms a major part of the Tanzanian landscape, spreading west from the Indian Ocean to cover one-fifth of the country's total land area. From the 1920s on, British colonial administrators had considered various plans to develop the waterway, but they generally sided against the construction of hydroelectric dams, taking into account the challenges of seasonal variations in water flow, the potential for unpredictable floods from upstream farming practices, and the need to protect turbines from sand and silt. Moreover, having worked closely with the local residents, these colonial administrators recognized that periodic flooding was

---

*S. A. Waldo, as quoted in David Ekbladh, *The Great American Mission: Modernization and the Construction of an American World Order* (Princeton: Princeton University Press, 2010), p. 102.

in fact precisely what made the region so agriculturally rich: the replenishing of alluvial soils made it possible for farmers to grow pumpkins, tomatoes, tobacco, sugarcane, maize, cotton, millet, and cowpeas without applying fertilizer.

Despite the sound advice of their predecessors, late colonial officials eventually fell under the spell of the TVA. In 1952 the colonial government partnered with the UN's Food and Agriculture Organization (FAO) to conduct a large-scale quantitative survey of the region, ultimately producing a seven-volume report on water flow, rainfall, topography, and soil type. Unlike previous surveys of the area's potential, the FAO reports were based largely on aerial photography and emphasized scientific information over the realities of local agricultural practices. When Julius Nyerere, Tanganyika's first president, began formulating his plans for development in 1961, these were the reports from which he worked. Like Nasser in Egypt, Nyerere was enthusiastic about the potential of hydroelectric dams to modernize his country. In the grandest plans, a giant dam across Stiegler's Gorge on the Rufiji River would produce upwards of 400 megawatts of electricity—more than enough to generate investment income for the nation by selling its excess power to neighboring countries in eastern Africa.

As both a committed socialist and a member of the Non-Aligned Movement, Nyerere attracted economic and technical assistance from sources as varied as Norway, Sweden, Japan, China, and the United States (and was given the obligatory USAID-sponsored tour of American dams). Nyerere envisioned the transformation of the Rufiji Valley as essential to his plans to transform Tanzania through collective farming. Some small dams, such as the gravity-driven Hale plant on the Wami River, opened as early as 1964. But while international funds and studies continued to support the Stiegler's Gorge project throughout the 1960s and early 1970s, lingering doubts about both the dams and collectivization stalled construction. Foreign consultants generally recommended construction, pointing to the possibilities for electrification, flood control, and agricultural irrigation. Local experts at the University of Dar es Salaam, however, generally advised against it, pointing to findings from the same sorts of field surveys and observations of farming practices that earlier colonial administrators had considered. In 1984, after nearly two decades of planning and controversy, the government put the project at least temporarily on hold.

The consequences from building other, smaller dams further upriver seem to have proven the local critics right. Built using data from the FAO Rufiji Survey, with assistance from the Swedish International Development Cooperation Agency, the Mtera Dam began operation in 1988. Within ten years, the dam had turned into a boondoggle: a variable water supply on the Great Ruaha River means unpredictable power production and electricity shortages in Dar es Salaam. Despite a number of more successful hydroelectric dam projects in New Zealand, France, Brazil, Portugal, and elsewhere, the challenges in constructing the Stiegler's Gorge dam are indicative of the difficulties that soured the international aid community on the potential of giant dams to transform local economies. Nevertheless, faced with an ongoing electricity crisis and the allure of clean, renewable energy, planners in Tanzania have once again turned their attention to Stiegler's Gorge. In December 2010, the country's minister of foreign affairs announced a partnership with a Brazilian firm to explore the possibility of building a 2,100-megawatt hydroelectric plant at the site, for possible completion by 2015.

Building Expertise in India

Indian Prime Minister Jawaharlal Nehru placed science and technology at the center of his vision for a newly independent India. Although the British had left their former colony with an extensive education system, their technical colleges and universities had focused on training Indians to be effective civil servants and colonial administrators, not scientific or technical innovators. Nehru therefore made revamping the Indian higher education system a priority. In a Cold War context that stressed the role of education in fostering development, he found ready partners from across the political spectrum. The first Indian Institute of Technology (IIT)—established in Kharagpur, near Calcutta, in 1951—was modeled after MIT and built with international support from the United States, the United Kingdom, the Soviet Union, and UNESCO. As India became a major player in the Non-Aligned Movement, Nehru saw advantages in having specific nations support individual technical universities. There followed the creation of four additional technical universities, each tied to patronage from a different national sponsor: IIT Bombay (Soviet Union, 1958); IIT Madras (West Germany, 1959); IIT Kanpur (United States, 1960); and IIT Delhi (United Kingdom, 1963).

Although all the IITs were theoretically modeled after MIT, only IIT Kanpur benefited from an American consulting team drawn from that very university. MIT at first resisted the Indian government's appeal for assistance, but university leaders agreed to take on a partnership after encouragement from their patrons at the Ford Foundation and a push from their policy-oriented colleagues at CENIS. Soon MIT's advisors were joined by experts from a consortium of eight other leading American engineering schools. The American plan for Kanpur was truly based on the U.S. model, with American textbooks, American (as opposed to British) exams, and an American-style program for faculty and graduate research. In hiring its initial faculty, IIT Kanpur continued this emphasis on foreign expertise, with two-thirds of the faculty members having been educated in the United States (mostly Indians living abroad). The $14.5 million in American aid, primarily from USAID, that flowed into the institute during its first ten years financed the innovations that made the university the most competitive and prestigious in India.

In 1972, U.S. support for IIT Kanpur ended abruptly, a casualty of shifting Cold War alliances. With hostilities increasing between India and Pakistan, India had signed a treaty of friendship with the Soviet Union; the United States in turn offered its support to Pakistan. By this point, IIT Kanpur had already become an elite technical institution no longer in need of American federal support. Under its first director, P. K. Kelkar (an electrical engineer educated at the University of Liverpool), IIT Kanpur had invested heavily in computers. An IBM 1620 was one of the first pieces of equipment installed on the campus; a few years later, the university's IBM 7044 machine was one of the most powerful computers on the subcontinent. Many of these machines were already out of date by the time they arrived on campus, but this potential roadblock seemed only to encourage IIT students to develop their programming skills. IIT faculty conducted computer-training seminars across India, serving as catalysts for the development of the nascent Indian high-tech industry.

The withdrawal of American government support therefore did little to derail either IIT Kanpur's young tradition of excellence or its cozy relationship with the American computer industry. Ironically, the university may have succeeded too well in establishing an American model on Indian shores. From the beginning, the best of IIT Kanpur's undergraduates preferred to complete their professional educations overseas. This trend accelerated with the relaxation of U.S. immigration quotas in the mid-1960s. All told, over half of IIT Kanpur graduates (approximately 125,000 students) now live abroad. Silicon Valley alone employs approximately 200,000 Indians, many of whom studied at IIT Kanpur or under someone who had. Rather than ending India's brain drain, the creation of an "Indian MIT" appeared at first to accelerate it. Nevertheless, the other half of IIT Kanpur alumni did stay in India and undoubtedly have done their part to transform the Indian economy. From this perspective, too, the IIT Kanpur experiment may have succeeded too well: few American economists in 1961 would have envisioned the competitive challenge the Indian computing industry would present to American information technology workers just fifty years later.

Building Health in Mexico

Perhaps the most ambitious of all postwar development projects was the attempt to eradicate malaria, a tropical and semitropical disease spread by the *Anopheles* mosquito. Malaria had been the subject of numerous and repeated international public health campaigns throughout the twentieth century, but until midcentury, efforts had focused on control, not eradication. In 1955, encouraged by the ready availability of modern pesticides and recent successes with antibiotics, member nations of the World Health Assembly voted to attempt to eliminate the disease from human populations. At least 600 million people were exposed to malaria annually; eradicating the disease would not only save countless lives but would also boost economic productivity in precisely the geographical areas most in need of stimulus. Despite initial estimated costs of $600 million over five to ten years, public health authorities anticipated that the economic benefits would greatly exceed the costs. This would be a collaborative effort, with technical and material support from the World Health Organization, UNICEF, and the U.S. International Cooperation Agency (the precursor agency to USAID). Participating nations would contribute to local costs.

The case of Mexico mirrors the fate of the larger program. At the time the effort was launched, malaria was that country's third leading cause of death, with an annual estimated cost of $160 million in lost productivity and unworked land. The existing Mexican malarial control program consisted of draining marshes; spraying mosquitoes' breeding grounds with insecticides; and supplying affected populations with antimalarial drugs, window screens, and netting. The more ambitious eradication program, in contrast, attempted to spray the walls of every rural Mexican residence with the pesticide DDT. The khaki-clad public health worker, atop a horse and armed with DDT-filled spray canisters, became a familiar sight in the hills of rural Mexico. The antimosquito campaign fit well with the Mexican government's stated policy of increasing its control over remote, mostly indigenous areas: in 1957 alone, nearly 4,000 workers treated more than 3 million homes. At least at first, Mexico's program seemed successful, and its leaders took on the responsibility of training public health workers from across Latin America. By 1960, malaria had nearly disappeared from Veracruz, Acapulco, Guadalajara, and the urban north; the overall malaria infection rate had dropped by 90 percent, from 40,591 cases in 1955 to 3,665 cases in 1960.

This success was, unfortunately, temporary. Malaria eradication proved a particularly elusive goal in Mexico's more tropical, indigenous regions, especially Chiapas and Oaxaca. In part, this was a reflection of an environmental reality: malaria is simply more difficult to eliminate in locations where it is a recurring, rather than sporadic, part of the public health landscape. Neighboring Guatemala's less successful campaign also contributed to the failure, as infected mosquitoes and migrants regularly carried the disease across the border. But cultural and economic factors were also at play. The top-down structure of the eradication program had few provisions for dealing with either native languages or locals' resistance to admitting strangers into their homes. Many rural Mexicans in the south either moved frequently or slept outdoors in hot weather, thereby missing out on the supposed benefits of a home sprayed with DDT. And with the emergence of DDT-resistant mosquitoes, sprayers switched to even more toxic pesticides that killed pets, livestock, and even occasionally people. International enthusiasm for purchasing these insecticides lagged, in part because of their toxicity and in part because the Mexican program increasingly used domestically produced insecticides. During the late 1960s and early 1970s, one international group after another withdrew its funds from the Mexican malarial eradication campaign, though the domestic program staggered along until the 1970s. By 1970, Mexican officials reported almost 50,000 cases of malaria scattered across nearly 60 percent of the country.

In 1969, the World Health Assembly acknowledged that the $1.4 billion spent on global malaria eradication had failed in the countries that needed it most: tropical Latin America, Africa, and Asia. The eighteen countries that did successfully eradicate malaria by 1970 tended to be small and isolated (islands like Taiwan, Saint Lucia, and Mauritius), socialist (Cuba, Bulgaria, and Yugoslavia), or economically developed (the United States, Australia, and the Netherlands). Most importantly, given the original arguments made on behalf of development, no one was able to produce conclusive data showing that malaria eradication brought economic benefits. Indeed, some critics (both then and now) argued that the primary beneficiaries of foreign aid for malaria eradication campaigns were American chemical manufacturers. In 1961, for example, USAID purchased more than one-third of all insecticides produced within the United States for use in the global campaign. Today, malaria—now often in drug-resistant form—kills an estimated 2 to 3 million people a year, mostly in Africa. Public health authorities and economists still agree that malaria and other tropical diseases are a drain on economic development, but they no longer propose massive top-down programs dependent on foreign aid and expertise as the solution.

Global enthusiasm for Third World development projects declined dramatically in the early 1970s, when the international aid community was shaken by accusations of corruption, political expediency, and covert ties to intelligence operations. Rather than transforming the Third World, a decade of development work seemed only to have demonstrated that neither foreign investment nor technical expertise would be sufficient to solve the problems of global poverty. Yet this bleak assessment is surely too glib: development funds built schools, hospitals, and water treatment facilities that almost certainly improved the lives of hundreds of millions of people. The difficulty in evaluating the legacy of the Development Decade has less to do with the specific projects undertaken by institutions and individuals promoting such varied agendas as socialism, global capitalism, African nationalism, or simply the idea of a better life, than it does with the problematic assumptions at its heart.

The tragedy of American foreign aid was that its architects too quickly forgot Walt Rostow's dictum that "an economy is an instrument for a larger purpose." In their fervent belief that global consumer capitalism would inevitably usher in an era of freedom and democracy, American foreign policy advisors all too often embraced partnerships with repressive regimes whose leaders endorsed their theories of economic growth. The parallel to Marxist philosophy, with its view of political change as the inevitable result of economic processes, was largely lost on these theorists. Nowhere did this tragedy have more visible results than in South Vietnam, where the Americans attempted to prop up the regime of dictator Ngo Dinh Diem by "modernizing" the countryside through democratic experiments, canal construction, and education—a process that also involved mandatory relocation, forced labor, and barbed wire enclosures. Even after the assassination of Diem in 1963, the U.S. approach to counterinsurgency continued to stress modernization over self-determination. Even as U.S. forces were in the process of dropping more bombs over North Vietnam than had been used in all of Europe during World War II, U.S. technical advisors promoted schemes to transform economic life in the south (and throughout all of Southeast Asia) through the construction of a giant hydroelectric dam on the Mekong River delta. Rostow, as Johnson's national security advisor, endorsed both the bombings and the dam.

The increasingly disastrous foreign policy consequences of uncritical development theory are only part of the explanation of why the broader project failed. "Development," as embraced by Americans, Soviets, and Third World leaders alike, assumed an unabashedly positive view of science and technology as forces for good in the world. Despite the ever-present fear of a nuclear holocaust, these theorists and policymakers overwhelmingly believed that the products and techniques of modern science, particularly state-sponsored science, would make life better. To understand why Americans began to reject this idea, we must turn our attention to the mixed results of applying Cold War–era science and technology to American life.

5 Science and the General Welfare

The early 1960s marked a high point for American faith in the power of science to solve problems at home and abroad. Americans placed their confidence not only in the physical sciences, on whose fate policymakers had pinned the national interest since World War II, but also in the social sciences. Economists, sociologists, psychologists, and even managers were increasingly called upon to help solve America's problems as well as explain its success. Meanwhile, advertisements, magazines, and films praised the wonders of modern science and technology that had made American consumers the envy of the world. The message was clear: Americans were affluent, and science had made them so.

This powerful image of a happy, prosperous midcentury America, fueled by advances in science and technology, obscured a more complicated reality. While lavish federal funding for science and technology had transformed the practice of American science, it had produced few of the wonders of consumer technology predicted in Vannevar Bush's *Science: The Endless Frontier*. Nor had prosperity reached all Americans equally. Poverty remained a fact of life, especially among African Americans who lived in increasingly segregated cities. As the 1950s turned into the 1960s and riots broke out in Watts, Newark, and Detroit, the problems of urban America became harder to ignore. At home, the urban crisis threatened unrest and economic instability; abroad, it undermined the nation's claims to moral leadership for the rest of the world.

In response, President Lyndon Johnson's "War on Poverty" proposed to stage a full-frontal assault on the United States' domestic ills. The military metaphor is apt: many of the Great Society's antipoverty programs were headed by social scientists who had cut their teeth on 1950s-era defense projects. Where once the social sciences had had to fight for inclusion in the NSF, now they led the battle against poverty, racism, political disenfranchisement, and urban decay. But America's social problems, it turned out, were not well suited to theories developed in defense laboratories. When the Great Society's social scientists failed to restore America's cities, faith in scientific thinking started to fail too.

The Promise and Perils of Affluence

Many Americans of the 1960s, including many economists, understood themselves to be living in the most modern and most successful country on earth, a place where freedom meant not only political rights but also the ability to buy space-age consumer goods. The postwar years were indeed prosperous: between 1949 and 1973, average family income doubled. As befit the world's leading capitalist nation, private spending fueled the boom, consistently providing about two-thirds of the nation's gross national product. Every new suburban home required a refrigerator, a stove, a furnace; the arrival of millions of baby boomers drove spending on clothes, furniture, and doctors' appointments. Newly available forms of credit, such as installment buying and department store credit cards, encouraged families to buy before they saved. And through it all, economists and politicians encouraged Americans to buy, buy, buy, depicting consumer spending as a patriotic duty that would drive economic growth.

Yet no one, aside from perhaps the most rabid anti-Communists, believed that the American economy was based purely on market principles and free enterprise. Postwar economists advocated a mixed economy that included government spending to drive full employment, consumer spending, and growth. Many of the new tract homes in the suburbs, for instance, were built with federal dollars from the Servicemen's Readjustment Act of 1944 (better known as the GI Bill). New Deal–era entitlement programs like Social Security and unemployment insurance ensured a basic level of subsistence (and therefore consumer spending) for the aged and out-of-work. Overseas, U.S. foreign policy programs such as the Marshall Plan and Point Four helped establish a friendly climate for American business interests. The U.S. economy might be based on free enterprise, but it was a form of free enterprise shaped by government support and regulation.

One of the most important areas of postwar government spending, particularly for the defense sector, was in science and technology. As we saw in chapter 2, the DOD, the AEC, and the individual armed services invested enormous sums in R&D in a variety of institutional settings—universities, national laboratories, nonprofit think tanks, and private industry. Defense contracts powered the aerospace, electronics, materials science, and computing industries. In the post-*Sputnik* era, the NSF channeled federal dollars into less obviously defense-related scientific work, including so-called basic research. Over the course of the 1960s, the NSF gradually welcomed the social sciences, too, into the fold. Scholarships and fellowships for future scientists and engineers represented yet another avenue for federal investment in science and technology. In 1964, federal spending on R&D, including basic research, reached a historical high of 2.88 percent of gross domestic product. Despite professional societies' consistent complaints of low pay for federal scientific and technical workers, a scientific career offered a reliable path to middle-class prosperity for hundreds of thousands of postwar chemists, physicists, psychologists, and engineers.

Aside from its potential to produce direct military applications, science was considered a good investment in the overall economy. Vannevar Bush's 1945 report to the president, *Science: The Endless Frontier*, had stressed the importance of basic research not only for national defense and the nation's public health, but also for driving the creation of civilian jobs. Writing only five years after the nation had started to pull itself out of the economic doldrums of the Great Depression, Bush forecast a future of full employment based on "new and better and cheaper products." But, as Bush reminded his audience, "new products and processes are not born full-grown. They are founded on new principles and new conceptions which in turn result from basic scientific research." Scientific research, as Bush described it, was a form of "scientific capital."* In the same way that the government might stimulate corporate investment through fiscal policy, it might encourage the development of new consumer goods by subsidizing scientific research, revising the tax code, and clarifying the patent process. Even though the eventual NSF bore little relationship to the "National Research Foundation" Bush had originally proposed, many science policymakers continued to hope that federal investment in science and technology would produce technological spin-offs for the civilian economy.

Yet by the 1960s, federally funded research had yielded few specific consumer innovations. The most iconic products of postwar mass consumption—cars, refrigerators, televisions and nylons—had all been invented prior to World War II. Computers were slowly working their way into business administration, but in 1960 they remained a specialized and arcane technology primarily used in military applications. Atomic energy, too, had fallen short

*Vannevar Bush, *Science: The Endless Frontier; A Report to the President* (Washington, DC: Government Printing Office, 1945). Available online at www.nsf.gov/od/lpa/nsf50/vbush1945 .htm.

of its promoters' promises to cure cancer and produce electricity too cheap to meter.

In the late 1950s, prominent economists and engineers began attempting to intervene in what they perceived as lurking threats to America's continued economic growth. Economist John Kenneth Galbraith's best-selling book *The Affluent Society* (1958) criticized the structural underpinnings of the American economy, questioning the sustainability of growth fueled by consumer debt and military spending. American society suffered from a glut of private goods and a dearth of public services, a situation Galbraith drove home with the evocative image of the family who drive to a picnic in their new "air-conditioned, power-steered, and power-braked" car, only to dine on the banks of a polluted stream in a park filled with criminals and vagrants.* Private manufacturers spent their research dollars solely on bells and whistles intended to stoke consumer desires, he argued, while federal investment steered innovation toward military goals. Similarly, a 1962 report by the Engineers Joint Council issued dire warnings about the "imbalance" created by overinvestment in defense, atomic energy, and space, to the neglect of urban infrastructure and environmental controls. The massive input of federal defense dollars into private research laboratories, these critics argued, was blinding American researchers to the diverse needs of the civilian economy.

News from Japan—specifically, reports of its red-hot consumer electronics industry—stirred worries about even those industries that flourished under military investment in R&D. The solid-state transistor, a device for controlling and amplifying electrical signals, had been invented at Bell Laboratories in New Jersey in 1947. In the United States, the firms that developed applications for the transistor, particularly Westinghouse and RCA, worked closely with military customers. By the late 1950s, defense contracts funded 75 percent of the research conducted at RCA Labs—research that often had little to do with the transistors being manufactured at RCA's own factories. Meanwhile, the company had turned to foreign patent licensing and technology training as a way to boost its bottom line. The Japanese electronics companies that purchased this technology had no domestic military audience, since the Japanese constitution adopted under the postwar American occupation prohibited Japan from establishing a conventional army. Thus, instead of developing components for ICBMs, the Japanese firms Hitachi, Toshiba, and Sony focused on parts for televisions and hi-fis. In 1959, the Japanese semiconductor

---

*John Kenneth Galbraith, *The Affluent Society* (Boston: Houghton Mifflin, 1958), p. 253.

industry became the largest in the world, surpassing even the American laboratories pumping out parts for the Cold War machine. In the process, they
terrified American economists.

By the early 1960s, then, there was growing unease about the social and
economic consequences of a postwar boom built on defense spending and consumer culture. And yet, most of the mainstream discussions of the problems
of affluence ignored the reality that, in fact, not everyone was prospering. An
important exception was Galbraith's *Affluent Society*, which drew attention
to the "islands" of poverty that remained stubbornly resistant to the broader
phenomenon of American economic growth. Four years later, Michael Harrington's *The Other America* (1962) made the same points more stridently, revealing the ravages of poverty in Appalachia, the South, and cities, and among
minorities, the aged, and the mentally ill. Estimates placed the number of the
truly poor—those unable to feed, clothe, and house their families—at somewhere between 10 and 20 percent of the population.

A disproportionate number of the poor were African American. With Jim
Crow laws still in place in much of the South (and unofficially in place in
much of the North), even those African Americans with the financial resources
to participate in postwar mass consumption found themselves excluded from
the marketplace. In the midst of a political culture in which belief in economic
equality ran deep, social scientists sought to explain these disparities in terms
of individual behaviors. The most optimistic among them hoped that removing legal barriers to economic opportunity would level the playing field for
blacks and other minorities. The more pessimistic pointed to an endless cycle
of economic exclusion, perpetuated by a "culture of poverty."

The Psychology of Race

The ongoing reality of legally sanctioned racism threatened to undermine the
United States' claims to international moral leadership in the postwar era. Aside
from the inherent injustice of segregation and racial violence, the so-called race
problem had become an international scandal in light of American attempts
to court Third World leaders—particularly leaders in Africa. Newspapers as
far and wide as the Nigerian *Daily Telegraph,* the *Fiji Times and Herald,* and
the Mexican *Excelsior* reported on lynchings in Mississippi and the school
segregation standoff in Little Rock, Arkansas. When American restaurants and
hotels in Maryland and the South refused service to visiting African diplomats,
State Department representatives realized they had a serious public relations
problem. The need to deny the Soviet Union an ideological hammer played a

large part in President Kennedy's, and later President Johnson's, efforts to pass civil rights legislation. According to a 1963 Harris Poll, even average Americans recognized the international importance of improving race relations, with more than three-quarters of respondents agreeing that racism and discrimination harmed foreign policy.

Social scientists agreed that the United States needed to do something to improve the lot of African Americans, but for somewhat different reasons. A growing consensus among sociologists and psychologists held that racism was psychologically damaging to both blacks and whites. Some went so far as to argue that prejudice made people more vulnerable to appeals from authoritarian leaders, an argument that carried special weight in the context of the Cold War. In *An American Dilemma* (1944), a massive volume that synthesized most of the existing literature on American race relations, Swedish polymath Gunnar Myrdal wrote that prejudice damaged the psyche of African Americans by destroying their feelings of self-worth; it damaged that of white Americans by creating an inner conflict between their sense of self as enlightened bearers of freedom and the reality of their status as moral oppressors. For Myrdal, racial conflict was both an internal and an external phenomenon that took place both within and between individuals.

An American Dilemma launched a postwar approach to understanding and overcoming discrimination that focused on changing individual behaviors. Like the economists who urged individual Americans to do their part in propping up the economy by indulging in mass consumption, psychologists and sociologists sought to cure the effects of racism by carrying out change at the level of the family and the individual. In their efforts to reform society, these experts were drawing on a long tradition in the social sciences, but their focus on individuals, rather than economic or political structures, reflected a dramatic shift from prewar explanatory models. This is not to say that postwar social scientists ignored the legal barriers to equality that continued to plague African Americans through the mid-1960s; quite the opposite. But even in the legal context, social scientists based their case on the effects of racism and segregation on individuals rather than the group. In part, this reflected a shift in institutional support: whereas prewar social scientists were most likely to receive funding through private philanthropic foundations, postwar researchers received support from the same sorts of federal sources that supported the physical and biological sciences. As a general rule, social scientific projects funded with federal dollars were not asked to investigate the institutional sources of racism, poverty, or, for that matter, most other social ills. Instead, they focused on ways to adapt individual behavior to a social and economic system that was assumed to be—if not perfect—certainly better than the alternatives.

Psychology, in particular, had flourished as a military science in World War II and the early Cold War. Psychologists offered advice to the Army on psychological warfare, intelligence techniques, classification schemes, and aptitude testing. A new national research organization, the National Institute of Mental Health, was founded in large part to investigate the root causes of the psychiatric disorders that plagued Army recruits. Problems of international development, too, could be considered through a psychological lens, as witnessed by Lucian Pye's theories of the appeal of Communism in Asia (see chapter 4). And once soldiers left the service, they carried with them the idea that clinical therapy—no longer the domain of the wealthy and the neurotic—was an appropriate and desirable treatment for unhappiness, discouragement, or stress.

It was only natural, then, that psychologists turned their newly recognized expertise to the problems of African Americans and those who persecuted them. Drawing on Myrdal's work, many studies in the 1940s and early 1950s assumed that whites' dedication to what Myrdal called the American Creed—a deeply held belief in equality, democracy, individual opportunity, and the rule of law—would eventually win out over racist attitudes. In Boston, for instance, a 1944 project attempted to reduce prejudice among police officers by allowing them to air their perceived grievances against black residents; a discussion leader then pointed out the frustration and traumas that drove black behavior. Civil Rights protests in the 1950s similarly appealed to white Americans' sense of justice and common decency. By the early 1960s, however, it had become clear to all involved that neither education nor awareness would be enough to overcome racial discrimination, particularly in the South. Researchers and policymakers therefore increasingly turned their attention to remediating the effects of racism in African Americans themselves.

Two basic and somewhat incompatible ideas drove federal antiracism measures in the 1950s and 1960s. On the one hand, social scientists believed that entrenched racism had produced a damaged "Negro personality" that featured slyness, self-hatred, passivity, immaturity, aggression, and neurosis—an attitude that famously found its way into a footnote of the U.S. Supreme Court's *Brown v. Board of Education* decision (1954) that declared school segregation a violation of the equal protection clause of the Fourteenth Amendment. The Court's decision referred to the "sense of inferiority" that inevitably results from segregated education. Reducing state-sanctioned discrimination was therefore a necessary first step in improving the economic situation of African

Americans by making them more confident, assertive, and ambitious. This viewpoint, considered liberal at the time, stood in marked contrast to biological explanations of racial disparities that had held sway for most of American history. African Americans weren't poor because they were hereditarily lazy or unintelligent; they were poor because they lacked self-esteem and motivation to achieve.

Lack of motivation, however, could be read less charitably. A second approach saw African Americans as mired in a "cycle of poverty" perpetuated by the supposedly pathological structure of the black family. Whatever the cause of blacks' poor economic attainment, this theory held, attitudes of hopelessness and low self-worth were being reproduced in future generations by a matriarchal family structure that denied masculine achievement. The matriarchal family thesis, largely drawn from the work of black sociologist E. Franklin Frazier (whose 1939 book *The Negro Family in the United States* put forth many of the same arguments that would frequently appear in 1960s-era discussions of race), pointed to the historical role of slavery in destabilizing the black family. As developed in the 1960s, however, the theory got tangled up in Cold War gender norms that insisted that a woman's primary role was in nurturing her children's psychological growth. From this perspective, a black woman who worked outside the home threatened to drive her children deeper into, rather than out of, the cycle of poverty.

This second, more controversial interpretation of the psychosocial causes of African American poverty reached its widest audience in the form of a policy planning document written for the Johnson administration. In early 1965, Assistant Secretary of Labor Daniel Patrick Moynihan was asked to prepare a background document on three decades' worth of social science research on race, poverty, and the family. His resulting study, *The Negro Family: The Case for National Action* (1965), summarized the by-now familiar themes of low self-esteem, dependence, and lack of male breadwinners. Like others before him, he presented the black family as unstable, damaged, and disorganized. What distinguished the so-called Moynihan Report from previous sociological accounts, however, was its relentless focus on the seemingly objective tools of social science, including more than two dozen tables, charts, and graphs that, one by one, showed correlations between divorce, out-of-wedlock births, male unemployment, and welfare payments. And perhaps more to the point for the report's eventual influence, the first media stories on its content appeared within a week of the Watts riots.

The riots that rocked Watts, a neighborhood of Los Angeles, in August 1965 were the largest and most destructive to hit the United States since World

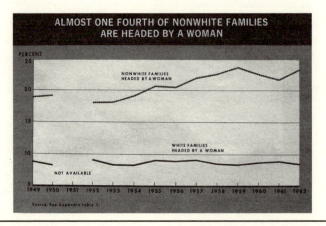

ALMOST ONE FOURTH OF NONWHITE FAMILIES ARE HEADED BY A WOMAN

PERCENT

NONWHITE FAMILIES HEADED BY A WOMAN

WHITE FAMILIES HEADED BY A WOMAN

NOT AVAILABLE

1949 1950 1951 1952 1953 1954 1955 1956 1957 1958 1959 1960 1961 1962

Source: See Appendix Table 5.

"Almost One Fourth of Nonwhite Families Are Headed by a Woman"

The Moynihan Report (1965) used the visual technologies of the social sciences, including charts, graphs, and tables, to argue that a matriarchal family structure contributed to black poverty. Drawing on the insights of a previous generation of black sociologists, Assistant Secretary of Labor Daniel Patrick Moynihan suggested that the combination of slavery and Jim Crow laws emasculated fathers, leaving mothers as the dominant authority figure. In urban America, black women's participation in the workforce (56 percent of black women age 25 to 64, compared with 42 percent for white women) perpetuated this trend. Many of the Great Society's human development programs were therefore designed to fill the voids supposedly created by the lack of an appropriately submissive and nurturing mother. Although intended as a sympathetic document, the report drew immediate fire from critics who accused Moynihan of blaming the victim and ignoring the structural causes of economic inequality.

■ Office of Policy Planning and Research, United States Department of Labor, *The Negro Family: The Case for National Action* (Washington, DC, 1965), p. 11.

War II. The arrest of an African American by a white police officer after a traffic incident sparked five days of looting, burning, and fighting that eventually left thirty-four people dead and inflicted more than $40 million in property damage. It was in this context that the Moynihan Report became public knowledge, sparking immediate controversy. Policymakers from the Democratic Party embraced it as a blueprint for fighting the problems of black America by providing social stability and jobs (for men). Black activists and advocates for the poor castigated it as a vicious attack on the mothers and grandmothers who worked long hours at low wages to feed their children. Conservative commentators questioned the ability of the government to fix what Moynihan had identified as a problem of the family. Meanwhile the question of whether racial prejudice caused social pathology, or vice versa, remained.

By 1965, most of the legal barriers to racial equality had been removed. With *Brown v. Board of Education*, the Civil Rights Act of 1964, and the Voting Rights Act of 1965, African Americans increasingly had the right, on paper at least, to go to the same schools, live in the same neighborhoods, and vote in the same elections as their white neighbors. Many of these laws had been partially inspired by psychological research that placed the blame for unequal socioeconomic attainment squarely on racial prejudice. But by the mid-1960s, few seriously believed that legal equality could eliminate racial prejudice or break the cycle of poverty. Despite the weakness of an explanatory framework based on individual psychology in the midst of so much obvious social injustice, psychological explanations prevailed through the end of the 1960s. In the Great Society programs, the Johnson administration would attempt to conquer poverty and inequality by combining this individualist perspective with the best social scientific knowledge the defense industry had to offer.

The Science of Decision Making

The extraordinary range of programs introduced under Johnson's Great Society initiative reshaped the scale of the federal government. Energized by a landslide victory in 1964, Johnson and a Democratic Congress passed a series of bills that increased federal funding for and oversight of health care, education, transportation, urban renewal, environmental protection, and civil rights. The centerpiece of these efforts was the Office of Economic Opportunity (OEO), a new billion-dollar agency that oversaw the creation of antipoverty measures aimed particularly at African Americans. For help in managing such a massive enterprise, the Johnson administration turned to social scientists who had developed management tools in the biggest government operation thus far: Cold War defense research.

In 1965, the Johnson administration announced that the Pentagon's new approach to decision making—known as the Planning, Programming, and Budgeting System (PPBS)—would now be used throughout all government agencies. PPBS was an innovation of Robert McNamara, Kennedy's secretary of defense and the former president of Ford Motor Company, who had in turn adapted it from practices developed at RAND. Enthusiastic about the new managerial techniques, McNamara filled out his staff at the DOD with analysts trained in RAND's quantitative tradition. As developed at RAND, the concept of "systems analysis" referred to a means of evaluating strategic alternatives in terms of objective criteria. Both "inputs" and "outputs" could be quantified; costs should take into account missed opportunities as well as goals achieved. The approach attempted to impose a sense of order on the chaotic process of military planning and procurement, forcing military leaders to specify, for example, the merits of competing weapons systems on the basis of cost as well as accuracy and payload. In effect, RAND had revolutionized the approach to answering the age-old question of how much bang for the buck. Though never popular with military brass, PPBS was credited with major cost savings at the Pentagon, and McNamara was widely hailed as a financial genius.

As an explanatory framework, systems analysis brought together a suite of concepts that made sense for defense policy and theoretical economics but proved more difficult to apply to social problems. Military leaders, for instance, were already familiar with the technique of converting specific goals into numerical expressions, having witnessed the success of British operations research in identifying bombing targets and establishing supply lines in World War II. The accompanying mathematical tool of linear programming allowed analysts to calculate the maximum possible gain ("maximum output") for the lowest possible cost ("minimum input"). In game theory—a technique pioneered at RAND—economists paired these approaches with the idea of humans as rational actors to create a powerful resource for simulating competitive behavior. Economists used game theory to predict the outcome of situations as critical as nuclear war and as mundane as the likelihood that someone would repay a loan. The notion of "human capital," developed in the late 1950s and early 1960s at the University of Chicago, provided another way to transform human behavior and skills into objective, quantifiable criteria: education, health, and training could be considered "inputs" that would increase an individual's return on investment—that is, his or her earning potential. Collectively, these econometric approaches informed both the content and administration of Great Society programs.

Of those Johnson appointed to administer his War on Poverty, most shared McNamara's enthusiasm for systems analysis. Few had much experience working with social programs. The first director of the OEO's Division of Research, Plans, Programs, and Evaluation (RPP&E), Joseph Kershaw, had formerly served as the research director for RAND. As its name suggests, the RPP&E was the OEO's equivalent of the Pentagon's PPBS, charged with long-term planning, cost-benefit analysis, and program evaluation. The first task at hand for Kershaw and his colleagues upon their arrival in the summer of 1965 was to make sense of the explosion of community-oriented programs that had been established immediately after passage of the Economic Opportunity Act in 1964. Under the charismatic leadership of Sargent Shriver, the former head of the Peace Corps and President Kennedy's brother-in-law, the OEO had spent more than $200 million in its first six months, primarily on innovative (and therefore experimental) programs aimed at direct engagement with the poor. Shriver had originally intended the Community Action Program to serve as a sort of antipoverty laboratory, but social science methods were largely set aside in the rush to show results. Some of these programs, such as Head Start and the Jobs Corps, would prove immensely popular, but others drew political fire for placing large sums of money in the hands of inexperienced, and sometimes radical, community leaders. The impression in 1965 was of an agency in the midst of uncontrolled growth, willing to fund almost anything that might plausibly combat poverty. Kershaw's job was to impose order on the system.

Kershaw and his staff did not lack for ambition: one of their first actions was to issue a five-year plan whose goal was the eradication of income-defined poverty by 1976. This was an objective that could be easily translated into numbers and formulae. The RPP&E adopted an elaborate definition of the poverty line that encompassed 124 variations on family situation; at any given moment, economists, statisticians, and demographers could use surveys and sampling methods to determine how much ground the government would need to cover to make its goal. *How* to eradicate poverty, on the other hand, was a substantially more difficult matter. The OEO's approach, as defined in the five-year plan, was to reduce or eliminate social and economic factors that prevented poor people from participating in the larger economy. The resulting strategy put into practice the economic theories that had grown out of postwar America's affluence: federal spending to drive overall growth, particularly through hiring in the public sector; investments in human capital, such as job training, community health centers, and early childhood development programs; and a universal income guarantee designed to ensure access to consumer goods. And because Kershaw and his colleagues realized it would be difficult to tell if any of these measures were working, they made the development of formal, quantitative program evaluation techniques a priority for the OEO.

The challenges inherent in conducting cost-benefit analyses of social programs are well illustrated by the example of Head Start. The national Head Start program grew out of a 1965 summer demonstration project intended to prepare low-income children for kindergarten. According to the "cycle of poverty" analysis, these children lacked the cognitive and psychosocial skills nurtured by middle-class parents. The program immediately proved a community and public relations hit; by the fall of 1965, Head Start had been expanded to

Head Start Classroom

Many of the programs established by
the Office of Economic Opportunity
attempted to remedy the psychosocial
effects of poverty and discrimination.
One of the most popular and long lasting was Head Start, a comprehensive
early childhood program that provided
education, nutrition, health services, and
parental counseling for low-income children. Unlike most OEO programs, Head
Start was immediately embraced both
by its political sponsors and those it was
designed to serve. The support of millions
of constituents made it nearly impossible
for President Richard Nixon to cancel the

program even after a social science–based
evaluation questioned its effectiveness.

■ Photograph by the Office of Economic
Opportunity; courtesy of the National
Archives

a preschool program than enrolled more than half a million low-income children. Enrollments would quadruple to two million by 1967.

The problem was that the OEO's own internal data cast doubt on the
program's effectiveness. Studies by educational psychologists showed that the
initial gains made by Head Start children faded out during the course of the
school year. In 1968, facing increased pressure to cut costs, the RPP&E issued
a research contract for a control-group study of Head Start's effectiveness that
focused solely on educational test scores instead of the broader health, nutrition, and psychological benefits claimed by the program's advocates. Much
public outrage ensued, and the new Nixon administration decided to keep
the program despite its supposedly mixed track record. It turns out that ad-
ministering popular social programs poses political as well as methodological
challenges.

Other attempts to import military techniques into War on Poverty programs yielded similarly mixed results. Robert C. Weaver, the first secretary of
the Department of Housing and Urban Development, encouraged planners in
his new department to experiment with a variety of techniques imported from
military agencies, including linear programming, computer simulations, and
aerial surveillance. In Los Angeles, SDC applied its computer programming
skills to simulate alternative approaches to combating blight. In New York,
Mayor John Lindsay invited RAND to create a new institute for predictive

modeling and resource management. Such partnerships were common in the late 1960s, with benefits for both the defense contractors, who diversified their markets, and the administrators, who gained scientific authority. But while these approaches undoubtedly improved city administration, it was unclear whether they did much to improve the lives of black urban residents, who continued to suffer from racism, subpar housing, limited economic opportunities, and geographic and cultural isolation. Moreover, the defense contractors' analytical skills tended to be accompanied by less welcome habits of secrecy and budget overruns. By the mid-1970s, the fad seemed to have run its course, with most cities severing their ties to defense experts.

With the election of Republican presidential candidate Richard Nixon in 1968, the Great Society programs lost their political champion. Though many of the antipoverty programs would continue to grow under Nixon and Gerald Ford, the OEO was eliminated in 1973. Historians and contemporary commentators have offered a number of explanations for the demise of the War on Poverty, ranging from its overly ambitious promises, to the escalating costs of the Vietnam War, to a general disenchantment with the principles of liberal government. As one American city after another erupted in riots in 1967, 1968, and 1969, conservatives increasingly embraced cultural explanations of racial inequality, arguing that no amount of federal funding could offset the pathologies reinforced by instability in the black family and reliance on federal welfare programs. Yet, while domestic policy planners' enthusiasm for a direct transfer of military planning techniques to social policy declined, they generally deemed the *concept* of cost-benefit analysis a success. By 1972, the government was investing more than $100 million annually in social science evaluation procedures. Poverty did not disappear, but the new methods of planning, budget forecasting, and program evaluation were here to stay.

Despite their best efforts, American economists, sociologists, and psychologists failed to eliminate poverty, racism, and urban decay. The very idea that they might have been successful bears witness to the enormous faith placed in scientific thinking in the Cold War. Science had won World War II; so far, despite a few close calls, it had held off the Communists. Game theory, linear programming, and computer simulations—all tools developed to address military problems—offered the most powerful means yet devised to solve social problems. Anything seemed possible. Why not apply those tools to the nation's most pressing needs?

This optimism proved short lived. A systems analysis approach to decision making can certainly help policymakers identify the best choice among limited options, but the results will only be as good as the assumptions embedded in the variables. A cost-benefit analysis might show, for instance, that job training programs offer a better return on investment than subsidized college loans for low-income high school graduates, but the model might not take into account discriminatory hiring practices, regional deindustrialization, or the educational aspirations of urban youth. In their focus on individuals, these models almost certainly left out the macroeconomic policies that had so worried Galbraith, Harrington, and other, more radical critics—namely, that an economy dedicated to private enterprise, driven by consumer spending and military procurement, virtually ensured socioeconomic inequality. As "objective" commentators, Cold War–era social scientists rarely commented on the political commitments implicit in their work.

The growing disillusionment with social scientific models was part of a larger cultural shift already under way by the mid-1960s, as dissenting voices increasingly criticized not only the fruits of science and technology, but also the underlying political assumptions that built Big Science. In 1962, the publication of Rachel Carson's *Silent Spring* awakened its middle-class readership to the environmental dangers of widespread pesticide use, a phenomenon that came to symbolize the dangers of science more generally. Two years later, the nomination of conservative Republication presidential candidate Barry Goldwater signaled growing doubts about the proper scale and scope of government. And slowly but surely, students and faculty at campuses across the nation began to revolt against the creeping influence of the military-industrial complex on higher education. Yet even as bombs fell on North Vietnam, protests erupted on campus, and despair threatened to engulf American cities, the scientific R&D establishment had one last feat in store for both Americans and citizens of the world: on July 20, 1969, NASA put a man on the Moon.

6 The Race to the Moon

When astronauts Neil Armstrong and Buzz Aldrin stepped onto the Moon on July 20, 1969, they gave the impression of being utterly alone. Here were two men standing on the surface of a rock 250,000 miles from all other humans, with one exception: the even lonelier Michael Collins, who circled above them in lunar orbit, safeguarding the craft that would take them home. Yet they were simultaneously at the center of an international spectacle. An estimated 600 million people—one-sixth of the world's population—watched Armstrong and Aldrin on live television, an event only recently made possible by communication satellites. And the three astronauts of *Apollo 11* could hardly have gotten to the Moon by themselves. The lunar landing capped a decade of engineering and managerial feats, an enterprise that at one point had employed more than 400,000 people and absorbed almost 4 percent of the entire federal budget. The nylon flag Armstrong and Aldrin planted may have been for "all humanity," but it bore the Stars and Stripes.

The relationships between science, technology, and national power built into the Apollo moon-landing program make it a quintessential Cold War phenomenon. In the quarter-century between the Manhattan Project and the lunar landing, policymakers and national security advisors agreed that the United States' international supremacy depended on unfettered federal access to the nation's scientific resources. This is not to say, however, that nothing had changed between 1945 and 1969. In the 1950s, the distinction between "military" and "civilian" was largely hypothetical as the need to maintain a nuclear lead trumped concerns about the negative effects of secrecy, technocracy, and federal priorities on scientific research. By the 1960s, the changing international course of the Cold War forced the Kennedy and Johnson administrations to pay as much attention to the *image* of American science and technology as to its actual products, a shift that had dramatic effects on the way the space program was structured and described. And at home, the growing visibility of economic inequality dictated that policymakers make a greater effort to connect scientific investment with domestic progress. In more ways than one, Apollo was the last hurrah of the Cold War military-academic-industrial complex.

The Peaceful Skies

From the vantage point of 1957, it was unclear what kind of place outer space would be. As an unoccupied and unexplored territory, space beckoned to both military commanders and sci-fi dreamers. The United States ultimately decided to position its manned space program in civilian terms, while the Soviet Union pursued a more military tack. The American strategy of "Open Skies," however, was based more on rhetoric than reality. Both American astronauts and Soviet cosmonauts flew into space atop repurposed military hardware, missiles that could carry bombs or reconnaissance satellites as easily as (or in fact more easily than) men. Moreover, an Open Skies policy would provide legal cover for overflight military reconnaissance, in contrast to the territorial laws that ruled the seas—hardly a civilian goal. Nevertheless, during the height of the space race the superpowers agreed that weapons had no place in space. Understanding the symbolic role of Apollo, a civilian program, therefore requires a brief detour into the military possibilities for space.

Military leaders in the United States and the Soviet Union had originally envisioned at least three separate approaches to space warfare. Traditional military strategists hoped to use long-range missiles to hurl weapons to targets hundreds or even thousands of miles from their launch sites. The higher the launch trajectory, the less likely that the missiles could be shot down by enemy forces. The notion of IRBMs and ICBMs grew directly out of German success with the terrifying V-2 rocket in World War II. Both the Americans and the Soviets spirited German V-2 scientists out of conquered territory in the months immediately following the fall of the Third Reich. The highest-ranking German rocket scientists, dreaming of the good life, made sure to surrender to the Americans. Relocated to the Army Ballistic Missile Agency in Huntsville, Alabama, Wernher von Braun and his technical staff worked feverishly to develop the Army's Redstone and Jupiter missiles. The Navy and the Air Force, meanwhile, supported work on their own missiles—Viking and Polaris (Navy) and Atlas and Thor (Air Force).

Even while developing the unmanned Atlas and Thor missiles, Air Force leaders—unlike the artillery enthusiasts in the Army and the Navy—assumed that war in space would require the presence of humans. Their marquee 1950s-era space programs focused on exploring the limits of speed and altitude for human flight. Chuck Yeager burst through the sound barrier in an Air Force X-1 in 1947; in the late 1950s and 1960s, the more advanced X-15 would send a series of test pilots fifty miles up, to the edge of the atmosphere. The Air Force imagined its pilots racing through space to intercept an enemy's incoming ballistic missile, or perhaps steering an American weapon toward its intended target. Other plans included placing humans in semipermanent orbit to provide a perch from which to launch weapons from space.

Yet despite the Air Force's insistence on human spaceflight, its advisors predicted what would come to be one of the most important military applications of space technology: reconnaissance satellites. In the mid-1950s, American intelligence had extremely limited means for determining the size or location of the Soviet arsenal, or even whether the Soviets had the weapons that they claimed. Fortunately, the ever-imaginative analysts at the RAND Corporation had an idea: satellites that could snap photographs deep over enemy airspace, unmolested and possibly even undetected, and well beyond the range of enemy guns. By 1957, the Air Force had contracted with Lockheed's Missile Systems Division to develop three separate possible systems for satellite reconnaissance: one based on simultaneous video transmission ("vidicon"); one that involved recovering film capsules dropped out of orbit; and one based on an infrared system that could scan the skies for missile activity. Until the Air Force could build and launch its high-tech spy cams atop its very own missile, it would rely on top-secret high-altitude (and highly illegal) flights over Soviet airspace to gather what intelligence it could.

By 1957, then, the Americans were making steady (if expensive) progress on a variety of defense technologies centered on space. But for all their foresight, American military leaders overlooked one critical aspect of competing with the Soviets in space: the psychological benefits of being first. Defense advisors familiar with the nascent American reconnaissance program dismissed the primitive communications capabilities carried aboard the Soviet *Sputnik* in October that year, but the damage was done. To the general public, it appeared that the Americans were hopelessly behind.

Nevertheless, being second had certain advantages. *Sputnik's* flight path established a legal precedent for outer space as international territory, clearing the way for the United States to put as many reconnaissance satellites into orbit as it could successfully launch. And while most American foreign policy advisors agreed that it would have been better to have been first, they also saw an opportunity to position the American space program as more open, peaceful, and cooperative than that of the Soviets. The Soviet launches, shrouded in secrecy, made no attempts to hide their military character. In a fitting paradox of Cold War logic, the American dedication to "openness" would pave the way for the spectacular success of the military satellite program in the early 1960s.

The most important step in defining the American space program as a peaceful one was the decision to make NASA a civilian agency. The National Aeronautics and Space Act of 1958, signed into law on July 29, 1958, gave NASA responsibility for all space activities, with the notable exception of weapons systems, military operations, and defense. NASA would be responsible for scientific research, spacecraft and instrumentation design, international cooperation, and manned spaceflight; the Pentagon's new research agency, ARPA, would maintain responsibility for reconnaissance, missile defense, and orbital weapons systems. In practice, this meant that NASA's purpose was some combination of basic science, prestige, and cover for military missions. Whatever the hoopla surrounding NASA's creation, the civilian agency's first budget was nearly 20 percent less than that same year's military space budget, at $242 million for NASA compared with ARPA's $294 million.

While PSAC and NASA administrators debated the agency's mission and goals in the late 1950s, military planners made steady progress on their reconnaissance satellites. After a series of failures, the Air Force recovered its first packet of film from orbit in August 1960. Launched under the cover name of "Discoverer 14" and supposedly a scientific research mission, the first successful Project Corona flight produced a twenty-pound roll of film that revealed airfields, surface-to-air missile sites, and rocket launch facilities throughout Eastern Europe and the Soviet Union. Within a few years, the United States had several satellites in geostationary orbit (meaning that they maintain a stationary position in relation to the ground) over the Soviet Union, while the Soviet Union had at least nine satellites in elliptical orbit over North America.

The images produced by reconnaissance satellites were some of the most closely held secrets of the Cold War: in the United States, the office created to manage this intelligence, the National Reconnaissance Office, was not even officially acknowledged until 1992. At the same time, the fact that both superpowers possessed such satellites was, and was intended to be, an open secret among diplomats and defense strategists. The theory of mutually assured destruction held that no nation was likely to launch a nuclear weapon in the face of immediate, and devastating, retaliatory attacks. Reconnaissance satellites were therefore seen as a military technology, yet a military technology that kept the peace. Hence, when Kennedy and Khrushchev signed the Partial Test Ban Treaty in 1963, they were willing to ban the testing of weapons in space but not to make it off-limits for military technologies. The idea of the peaceful, but not necessarily demilitarized, skies would profoundly shape the space race.

Destination Moon

For President Eisenhower, the space race was a sideshow that distracted from the larger objective of countering Soviet military capability. A fiscal conservative, Eisenhower balked at NASA's projected costs for various manned missions, including up to $40 billion for a trip to the Moon. It was only with the election of President Kennedy in 1960 that the prestige value of space spectaculars began to outweigh their spectacular costs. With a coterie of advisors who believed in the power of science and technology to transform society, both at home and abroad (see chapters 4 and 5), Kennedy reached a decision quickly. On May 25, 1961, just four months into his presidency, Kennedy made his famous announcement: "I believe that this nation should commit itself to achieving the goal, before this decade is out, of landing a man on the Moon and returning him safely to Earth." More than perhaps any other moment in the 1960s, this announcement proclaimed the increasingly central role of the image of science and technology in waging the Cold War.

Kennedy had not given the question of space much attention during his campaign, aside from pointing out how the United States seemed to be falling further and further behind the Soviets. Americans were playing a perpetual game of catch-up in the years immediately following *Sputnik*: in September 1959, the Soviets crashed a probe, *Luna II*, within 90 seconds and 270 miles of its target on the Moon; in August 1960, they recovered safe from orbit a menagerie that included two dogs, a rabbit, forty-two mice, two rats, and assorted fruit flies. The public face of the American space program meanwhile produced a notable string of failures, including multiple rocket explosions that advertised potential weaknesses with the ICBM program. In contrast, the part of the program that was going well—reconnaissance—was entirely secret.

Space exploration seemed a natural direction for the new administration, given Kennedy's interest in the United States' international scientific reputation and the longstanding interest of his vice president, Lyndon Johnson, in the topic (Johnson had shepherded the National Aeronautics and Space Act through the Senate). Yet Kennedy's science advisors, like Eisenhower's before them, advised against a manned space program. If scientific achievement were in fact the goal, they argued, NASA would be better off investing in robots and unmanned probes. The only interesting questions that required human spaceflight concerned physiology—an observation that highlighted just how risky such an enterprise might be. In contrast, they argued, the first U.S. satellite, *Explorer I*, had scored an impressive scientific achievement by revealing the existence of the Van Allen radiation belt, and it had done so without putting lives unnecessarily at risk.

The president's science advisor, Jerome Wiesner, did concede that an impressive and ambitious manned space program might benefit American national prestige. Secretary of Defense Robert McNamara and James Webb, NASA's new top administrator, agreed. With Kennedy focused on finding some aspect of space competition "which we could win," a Moon project began to sound more attractive. It fit the requirements of a project that was ambitious, inspirational, and technically difficult. Webb, an avid proponent of the new managerial techniques, argued that the difficulty of the enterprise was itself the point: a nation that could master the technical and administrative challenges of bringing back a man from the Moon would be a nation to emulate. McNamara hoped that turning over manned space exploration to NASA might rein in the Air Force. Johnson, for his part, embraced the federal spending inherent in a lunar mission as a means to kick-start the economy of the South and the West: this was a vision of Apollo as a TVA in the sky.

Despite lobbying from Johnson and McNamara, Kennedy remained skeptical. A PPBS-style cost-benefit analysis prepared by the Bureau of the Budget put the price tag at an estimated $20 billion. If the primary point of spending such an enormous sum was to impress Third World nations, would it not be better, Kennedy asked, to tackle something more directly relevant to their needs, like ocean desalination? Webb, too, cautioned against the dangers of making technical promises that couldn't be kept. The only thing worse than being the second nation to put a man on the Moon would be to kill a series of charismatic astronauts in the process.

Three events in April and May 1961 tipped the balance: two American setbacks and one accomplishment. On April 12, the Soviet Union lapped the Americans once again when cosmonaut Yuri Gagarin became the first man to orbit Earth. His safe return, covered with much enthusiasm in the international press, demonstrated conclusively that humans could survive conditions in space. The following week, CIA-trained Cuban exiles landed in the Bay of Pigs in what turned out to be a disastrous attempt to undermine Castro's government. By most accounts, this was the low point of Kennedy's presidency. But on May 5, 1961, Kennedy got a reprieve in the form of astronaut Alan Shepard's fifteen-minute suborbital flight, the first manned mission of the Mercury program. In keeping with NASA's culture of openness, the launch and recovery were broadcast on live television.

Compared with Gagarin's orbit, Shepard's up-down flight offered few "firsts." Nevertheless, the decision to broadcast the entire event yielded certain

Astronaut Alan Shepard

This photograph of Alan Shepard, the first American in space, was taken during a flight simulation prior to his voyage on *Freedom 7* on May 5, 1961. NASA's earliest astronauts all shared a background as military test pilots, a demanding and dangerous job that required them to fly new or experimental aircraft to the limits of their endurance. An important part of their task was to provide aircraft designers with feedback on the layout of the cockpit and controls. Since real test runs were not possible for the manned space program, astronauts spent countless hours in simulations. The simulations served not only as training runs, but also as essential feedback for the contractors who built the machines. It was at the astronauts' insistence that the Mercury capsules included not only an escape hatch, but also windows and manual controls.

■ Courtesy of NASA

rewards. The U.S. Information Agency reported that international media were impressed with the "openness" of the American program. Moreover, the recovery of Shepard's *Freedom 7* capsule from the Atlantic Ocean demonstrated that the Americans could direct their missiles with at least some degree of accuracy. And in what would become a central theme of the space race, media reports played up the idea that Shepard was not merely a passenger but, rather, an active participant in his flight. Unlike the Soviet cosmonauts, so the story went, American astronauts were no mere slaves to technology.

With the success of the first U.S. spaceflight, the die was cast. Three weeks later, Kennedy announced that Americans would attempt to go to the Moon. Congress assented, doubling NASA's budget for fiscal year 1962. And although the rest of Kennedy's speech is rarely recalled, the context is important: Kennedy discussed plans for foreign aid, economic recovery, disarmament, and closer partnerships with U.S. allies. Apollo was at the center of a grand plan to win the Cold War through peace, prosperity, partnerships, and propaganda.

Space Age Management

James Webb was right to warn of the technical and managerial hurdles to Apollo. Even today, the concept of a lunar landing seems impossibly daunting: as of this writing, no human has left Earth's orbit since 1972. Getting to the Moon and back required not only solving unprecedented engineering challenges for environments that could not be replicated on Earth, but also the coordination of thousands of components and contractors, all under the bright glare of the public spotlight.

From the beginning, policymakers knew that Apollo would require a massive mobilization of scientific and managerial resources. For some Kennedy advisors, including Johnson and Webb, this was part of the appeal. During the first half of the 1960s, Apollo provided business to more than 20,000 contractors in aerospace, electronics, and materials science engineering. To make sure that at least some of this federal research investment spurred private entrepreneurship, Congress instituted a more liberal patent policy that allowed NASA, at the administrator's discretion, to place inventions in the public domain. The agency invested in new R&D facilities in previously underdeveloped areas of the country, particularly an array of facilities arranged in a giant arc above the Gulf of Mexico. Universities also benefited. Instead of continuing the 1950s-era federal research contract funding patterns that favored such elite institutions as MIT, Stanford, and Caltech, Webb followed the recommendations of the Seaborg Report (see chapter 3) in using NASA dollars to increase the number of first-rate engineering institutions, especially in the South and the West. Apollo was the largest peacetime domestic public works project the United States had ever seen.

The managerial difficulties of the Apollo program extended beyond scale to the complexity and novelty of the entire system. Designing rockets and missiles is an inherently tricky business that depends on harnessing the explosive power of dangerous liquid or solid fuels under controlled conditions. Besides the risk of fire and explosion, the electrical and mechanical components aboard any spacecraft are subject to severe vibrations, changes in temperature and pressure, and the effects of previously uncharted electromagnetic fields. In zero gravity, a single loose bolt or metal filing could cause an electrical short, tens of thousands of miles away from the ability to make repairs. Moreover, both manned and unmanned spacecraft require automated navigational systems because the mechanics of flight in a vacuum render human reaction times moot. Automation would require not only miniaturized computer systems—a novelty at the time—but also communication systems sturdy enough to send and receive signals across unimaginable distances. And so on.

Humans, whether passengers or pilots, presented their own engineering challenges. For one thing, they required air. The familiar terrestrial mixture of nitrogen and oxygen is difficult to maintain in an artificial environment, but the alternative—filling the cabin with pure oxygen—presents a fire hazard. Bodies, like the instrumentation, would have to be protected from changes in temperature and pressure. For long journeys, provisions would have to be made for the astronauts to eat, drink, sleep, and void wastes. And then there was the problem of human psychology: how would individual astronauts respond in the moment to the terrors of the unknown? Encased in their capsules and spacesuits, threatened by their own breath, studded with biomedical sensors, the astronauts became part of the machinery of Apollo.

Even the choice of route posed uncertainty. NASA's first plans for a Moon landing echoed an image depicted in countless sci-fi films: a giant rocket would propel a spacecraft to the Moon's surface, carrying with it sufficient fuel and heat protection for the return trip. Planners soon rejected this so-called direct ascent method because of the enormous size of a rocket needed to lift the projected 75-ton payload out of the atmosphere. Instead, NASA's engineers considered various rendezvous scenarios, all of which took advantage of the Moon's low gravity and lack of an atmosphere to reduce payload weight. In the end, they selected the controversial notion of a "lunar orbit rendezvous" approach, in which a single multistage rocket carried two separate components to lunar orbit. At that point, the so-called lunar excursion module (LEM) would separate from the main capsule (command/service module, or CSM) for a short drop to the Moon's surface. The small, lightweight LEM would then rejoin the CSM in lunar orbit, where it would be abandoned once the lunar astronauts had returned to the main cabin. Experts agreed that lunar rendezvous offered the best chances for beating the Soviets and meeting Kennedy's end-of-decade deadline, but it was risky: safety estimates put the chances of returning all crew members back to Earth at only 85 percent.

Given the inherent complexity of the enterprise, prosaic questions of scheduling, procurement, and organizational oversight took on monumental importance. A change in the size of a socket at the intersection of a rocket and crew module, for example, could disrupt work at hundreds of points down the line. Cost overruns were rampant, particularly in the early days of the space race, when political enthusiasm for manned space exploration ran high. One of the most notorious examples was the Mercury capsule, which had been contracted to McDonnell in a cost-plus-fixed-fee agreement for $19 million in 1959 but ultimately cost $143 million by the time the project concluded in 1963. As the reality of the cost of sending a man to the Moon sank in, and Johnson and Congress turned their attention to funding the Great Society and the war in Vietnam, budgets inevitably tightened.

By forcing Webb and his senior managers to work within a budget, Congress was effectively asking them to clarify NASA's mission objectives, including its relationship to larger national goals. Was the Apollo mission essential to national security, meriting unlimited resources in a crash program akin to the Manhattan Project? The introduction of the PPBS to the Pentagon suggested that even the military had to justify its investments in science and technology; presumably, NASA would be subject to at least as much scrutiny. Should cost-benefit analyses compare Apollo's potential accomplishments with those of USAID, or with the OEO? Should the scientific goals of individual Apollo missions, supposedly critical to NASA's image as a peaceful civilian institution, be sacrificed in the name of reducing payload, and therefore cost? What might constitute reasonable sacrifices for safety in an inherently risky enterprise?

Webb's response to these essentially political questions was to reflect on Apollo as a model for the management of large systems. NASA's success or failure would not be determined by technical decisions, or even by modern techniques of systems analysis or systems management. Instead, its triumphs would rest in the organization's ability to respond quickly to technological and political feedback, "the capacity to adjust to and to move forward in an unpredictable and sometimes turbulent environment."* In his classic *Space Age Management*, published just before *Apollo 11*'s triumph and shortly after his retirement from NASA, Webb compared the social, political, and technical challenges of the lunar landing to other problems requiring "large-scale efforts," including pollution, energy, public health, and income inequality. "The great issue of this age," he wrote, "is whether the United States can . . . organize the development and use of advanced technology as effectively for its goals as can the Soviet Union, with its totalitarian system of allocating and utilizing human and material resources."† The point of Apollo would be to demonstrate, once and for all, that technological achievement was just as possible in a liberal democracy as it was in a command economy.

NASA's managers developed a number of techniques to ensure the success of Apollo in the midst of shifting political winds. The key to the enter-

---

*James Webb, *Space Age Management: The Large-Scale Approach* (New York: McGraw Hill, 1969), pp. 8–9.
    †Ibid., p. 17.

prise, they realized, would be in integrating the incredible range of components for the system, each built by a different contractor. Thus, in addition to the contractors responsible for research, design, and manufacturing, NASA hired engineering firms that specialized in making sure the parts and schedules fit. In the apt phrase of one historian, NASA effectively became a "bureaucracy of innovation" that combined checklists, feedback loops, protocols for changing configurations, simulations, and genuine engineering and scientific breakthroughs.

Perhaps inevitably, there were setbacks, none more disheartening than the deaths in January 1967 of astronauts Gus Grissom, Ed White, and Roger Chaffee in a fire during routine testing of the first Apollo craft. The Apollo fire and the subsequent congressional investigation challenged NASA managers' sense of themselves as careful decision makers who exposed their crews to calculated risks. The most damning findings questioned NASA's ability to identify and resolve safety issues given contractors' financial interest in telling the agency what it wanted to hear. Nevertheless, the death of Soviet cosmonaut Vladimir Komarov on *Soyuz 1* a few months later served to remind all involved that space exploration was, by definition, an inherently dangerous enterprise. Indeed, the more surprising thing is how very few astronauts, cosmonauts, and industrial contractors died in the process of launching humans on top of missiles.

Despite all the odds, Apollo was successful in its mission of placing an American on the Moon within a decade of the first man in space. Historians continue to debate whether the lunar landing represented a wise investment of American resources, but its impressiveness as a technological achievement has never been in doubt. Moreover, Apollo was operated as a civilian and highly public program that offered an alternative to the unabashedly military, and secret, focus of 1950s-era federal R&D. The American flag that Armstrong and Aldrin planted on the Moon was intended to broadcast American leadership through inspiration, not military domination. There were, however, limits to the lengths that Americans would go to demonstrate their commitment to openness in space.

International Cooperation: An Alternative Path?

Even as Kennedy and his advisors were calculating the political costs and benefits of a race to the Moon, they discussed using space as a venue to demonstrate U.S. commitment to international partnerships, particularly international scientific cooperation. As early as June 1961, Kennedy and Khrushchev went so far as to consider the possibility of joining forces in their attempts to send a man to the Moon. A bilateral lunar mission was not to be, but the 1960s did see the development of genuine cooperation in space, particularly in fields that posed little threat to potential military operations.

The initial push to put up satellites had taken place within the framework (some might say the cover) of the International Geophysical Year (IGY), an eighteen-month-long cooperative scientific exercise that brought together scientists, engineers, and amateur space enthusiasts from sixty-seven nations. From July 1, 1957, to December 31, 1958, scientists collected and shared data on the atmosphere, the oceans, the polar regions, solar activity, and the Earth's magnetic fields. Although satellite launches had not been part of the original plans for the IGY, by 1955 both the United States and the Soviet Union had announced plans to place a satellite in orbit. Given the timing of the IGY, and the United States' concurrent plans to develop a reconnaissance satellite, it is almost certain that American defense advisors saw the scientific program as a convenient excuse to develop multipurpose satellite launch capabilities. Nevertheless, early satellites did contribute to scientific knowledge, most notably in the discovery of the Van Allen radiation belts. Information on satellite flight paths, collected by volunteer "Moonwatchers" across the globe, significantly improved scientists' understanding of the Earth's size and shape.

The IGY established a pattern that would hold for space science throughout most of the 1960s. The United States, and, to a lesser extent, the Soviet Union, were willing to share data and research findings with their international colleagues but drew the line at hardware or engineering specifications. International cooperation was somewhat easier for the Americans than for the Soviets, in part because the formal separation of the U.S. space program into military and civilian divisions allowed NASA to build partnerships—albeit on American terms—with researchers worldwide. NASA's international affairs division arranged for international experiments, satellite launches, and conferences as well as short- and long-term researcher exchanges. Thanks in large part to these efforts, a genuinely international space community developed over the course of the 1960s, with long-lasting friendships created in fields ranging from astronomy and exobiology to geology and chemistry.

Outside of these areas, however, cooperation was limited. The motivations for and restrictions on American cooperation in space technology are well illustrated by two incidents from the mid-1960s: the establishment of an international satellite communications network and the development of European launch vehicles. As early as the 1950s, U.S. corporate communications giant AT&T had been exploring the use of satellites to extend its reach into global markets. Using private funds, AT&T's efforts focused on a relatively low-tech design: a reflective, nonsynchronous satellite that would require elab-

Earthrise

Astronaut William Anders took this iconic photograph, entitled *Earthrise*, while orbiting the Moon as part of the crew of *Apollo 8*. NASA had supplied the crew with a Hasselblad handheld camera and 335 available frames to record mission-specific images, including possible lunar landing sites, the performance of separating rocket stages, and characteristics of the Earth's atmosphere. About 20 percent of the frames, however, were reserved for "crew observation." Although *Earthrise* was not the first photograph of the entire planet—an unmanned vessel, Lunar Orbiter I, captured that honor in 1966—its stark beauty and the human drama behind its creation captured viewers' imagination. *Earthrise* became a touchstone of the environmental movement, a symbol of the lonely, fragile Earth floating on a sea of nothingness.

■ Courtesy of NASA

orate ground stations to decipher its signal. With the growing importance of the Third World in Cold War politics in the 1960s, however, policymakers feared that a private international communications network would inevitably exclude less developed nations. NASA and the DOD therefore sponsored research at Hughes Aircraft to develop a geosynchronous communications satellite that would require only minimal ground support—thereby putting global communications within the reach of the poorest countries, and the poorest countries within the reach of American television and radio.

When the International Telecommunications Satellite Consortium, or Intelsat, was created in 1964 to oversee the development of a global satellite system, the United States was guaranteed not only a majority vote, but also a management role in the form of Comsat, a quasi-private company created by Congress in 1962. The fifty ground stations in twenty-eight countries that broadcast the lunar landing drew their signal from Intelsat satellites that were owned by Comsat and built by Hughes. In countries where private interests were slow to build communications infrastructure, U.S. development dollars filled the gap. By the time that the Soviet Union managed to establish its own global satellite communications network in the 1970s, it had difficulty convincing all but its closest allies in Eastern Europe to join. In this case, the United States used international cooperation to draw allies (and potential allies) close while maintaining a managerial framework that ensured U.S. financial control.

A similar dynamic was at work in American attempts to encourage the development of a European launch vehicle, with a strategic twist. Since the early 1960s the European Launcher Development Organization (ELDO) had been struggling in its efforts to unite separate British, French, and German rocket stages with an Italian satellite. By 1966, ELDO was bogged down by problems with integrating the different components (an issue NASA was deeply familiar with), and both France and Great Britain were threatening to pull out. It was at this point that U.S. national security advisor Walt Rostow recommended that the United States offer technical assistance, including hardware, technical documentation, and managerial expertise, to its European allies.

This departure from previous practice was justified, Rostow and his counterparts at the State Department argued, because ELDO's potential disintegration would be a public signal of broader troubles within the European alliance. Moreover, technical assistance might deflect increasing complaints in European quarters of a "technological gap" between the United States and Europe. But perhaps most importantly, Rostow believed that assisting ELDO would divert European, and particularly French, resources away from more belligerent ends. In November 1965 France had successfully launched a satellite atop an advanced three-stage rocket, *Diamant-A*, that could presumably be developed into a weapons delivery system; any money that France invested in a cooperative rocket launcher would be money not invested in French ICBMs. In the end, however, the plan to build up ELDO fell through, largely because of American reluctance to do anything that might threaten U.S. leadership (both economic and political) of Intelsat.

The idea of cooperation in space was, then, limited from the outset to areas that posed little threat to American economic or military dominance. It was thus a fitting epilogue to the space race that the last Apollo mission was a symbolic gesture of friendship between the United States and the Soviet Union. The brief flower of détente in the early 1970s had given rise to the idea of expanded U.S.-Soviet scientific cooperation, culminating in a joint space mission. After two years spent designing an elaborate collar that could connect the two nations' previously incompatible systems, the 1975 Apollo-Soyuz Test Project brought together three astronauts and two cosmonauts for "handshakes in space." Beyond demonstrating that cooperation in space was possible, however, the voyage did little to further international scientific exchange or to ease Cold War tensions.

The great irony was that even at the moment that the United States and the Soviet Union had fully embraced the publicity value of international co-operation in space, both nations had already begun to turn toward a more aggressive use of space technologies. The decision in 1972 to develop the space shuttle as the successor to Apollo was made with the assumption that the new system could double as a conveyance for military space devices. In exchange for NASA's agreeing to lift its spy satellites into orbit, the Air Force agreed to stop developing its own launch vehicles. The Soviet Union, meanwhile, had begun to develop antisatellite technologies to counter powerful American space reconnaissance devices. With the Apollo-Soyuz Test Project, the space race had officially ended—but so had the commitment to space exploration as a peaceful, open civilian enterprise. In the waning days of the Cold War, science would once again become more important as a tool than as a symbol.

For a brief moment, *Apollo 11*'s descent to the Moon rekindled the fire of technological optimism. Looking ahead to the year 2000, Pan Am started taking reservations for commercial Moon flights. Wernher Von Braun predicted a Moon base, a Mars landing, and possibly even manned flights to the outer planets. People everywhere, across the globe, paused for a moment to reconsider the limits of human accomplishment.

Yet by 1969, this sense of technological can-do and its accompanying faith in Big Science had already become historical anachronisms. As had been predicted by skeptics all along, American citizens and policymakers soon lost interest in Apollo after the initial Moon landing. *Apollo 13* nearly ended in tragedy, but the other missions to the Moon—there were six successful manned landings in all—became routine. President Nixon canceled three additional planned missions and slashed NASA's budget to one-fourth of its Apollo peak; employment in the civilian space sector plummeted from its high of 420,000 in 1966 to 190,000 in 1970. Even before *Apollo 11*, social activists portrayed NASA and the lunar program as a distraction from the United States' limited achievements in conquering poverty, racial inequality, and urban violence, calling it a "Moondoggle," or worse.

As with U.S.-sponsored international aid programs and Johnson's Great Society, Apollo was initiated at a moment when most Americans, in and out of government, believed in the centrality of science and technology to the national purpose. The products of science and technology could inspire global friendship, improve the quality of life, protect against illness, yield a return on investment, and, above all, protect national security. Scientists had a special role, and a special responsibility, in ensuring that government tended and used this resource wisely. Compared with plans to eliminate poverty or develop the

Third World, Apollo was an unabashed success, but the world in which it ended was dramatically different from the one in which it had begun. However fragile or fraught the Cold War consensus on the role of science might once have been, by 1969 it was officially broken.

7 The End of Consensus

Despite a number of close calls, the United States and the Soviet Union avoided direct military confrontation during the Cold War. Instead, the two countries waged war on ideological and proxy battlefields. Some of these campaigns, like the Apollo missions, were intended to inspire global awe and inspiration. Others, like the war in Vietnam, attempted to accomplish foreign policy goals through brute force. As the death toll from Vietnam mounted, the gruesome consequences of the partnership between American science, technology, and the military became harder to ignore for both scientists and the general public. Over 58,000 Americans lost their lives in the Vietnam War; the number of North and South Vietnamese killed is harder to pinpoint but almost certainly exceeded two million.

Such devastating losses forced Americans to reconsider the myriad ways that military and civilian life had become intertwined. For scientists, the visibility of the war highlighted the difficulty of conducting scientific research while opposing military efforts. Such complaints were not new, but they gained a new currency in the wave of protests that swept college campuses in the late 1960s. By the early 1970s, the notion of science itself had become politicized as radical critics demanded that both public and private laboratories address environmental pollution, socioeconomic inequality, and the energy crisis. The consensus that had envisioned an obvious and unbreakable link between science, national security, and national prestige no longer held sway. In a remarkably short time, Vietnam sundered the foundations on which the expansion of postwar American science had been built.

Political Fallout

Accounts of American popular culture of the 1960s usually depict hippies, not scientists, as protesting the military-industrial complex. But this view is mistaken. Scientists and engineers participated in the 1960s-era critiques of what became known as "the system" with great vigor; some of the key texts were actually written by engineers. It is nevertheless true that public critiques of science, *as science*, were a relatively new phenomenon in the 1960s. Under-standing how members of the general public came to question the cultural authority of scientists requires a brief foray into an earlier protest movement, the protests against the dangers of atomic fallout in the late 1950s. Geneticists, in particular, became outspoken critics of national defense policy as growing concerns about the biological effects of atomic fallout sparked a nuclear test ban movement.

As we saw in chapter 2, those scientists who opposed the easy alliance of scientific research and state power in the early 1950s had few options for ex-pressing themselves; many of those who spoke up (for instance, Edward Con-don and J. Robert Oppenheimer) often had difficulty obtaining the necessary security clearances to work in their chosen fields. Less noteworthy scientists in political hot water often lost their jobs or were told that their contracts or research grants had not been renewed. Nevertheless, some scientists managed to carve out a space for dissent as individuals within the Cold War university and in the broader public sphere, particularly after Senator Joseph McCarthy's fall from grace in 1954. By the late 1950s and early 1960s, these fissures had started to spread to the formal institutions of science.

The logical starting point for opposition to defense policy might have been the Federation of American Scientists (FAS), the successor group to the atomic scientists' movement that had briefly thrived after the Manhattan Project. The members of the FAS had amassed significant experience in both public rela-tions and formal lobbying through their efforts to establish the AEC as a civilian agency. The FAS was loosely affiliated with the *Bulletin of the Atomic Scientists*, a magazine that became famous for its so-called Doomsday Clock, a graphical illustration of how close the world was to nuclear destruction. Throughout the 1950s, the dreadful hands of the clock marked seven, three, or even two minutes to midnight as the United States and the Soviet Union con-tinued to announce new and more powerful weapons. Scientists viewed the magazine as a forum where like-minded experts—originally physicists, but, starting in the early 1950s, increasingly biologists and social scientists—could debate the best strategies for bringing an end to the arms race and establishing a lasting peace. But these critiques had their limits: many of the contribu-tors to the *Bulletin* continued to pursue research projects funded by defense and quasi-defense agencies such as the AEC. These commentaries typically assumed that those scientists who contributed to defense technologies were best qualified to comment on their potential use.

The assumption that scientists associated with the defense industry pos-sessed the most authority to critique nuclear strategy gradually changed over the 1950s, in large part due to missteps by AEC chairman Lewis Strauss. On March 1, 1954, the AEC had detonated one of the first hydrogen bombs, designated Bravo, on the Pacific island of Namu, in the Bikini Atoll. The combination of shifting winds and an unexpectedly large explosion exposed at least 300 people—including American sailors and weather researchers, natives of the Marshall Islands, and Japanese fishermen—to significant levels of radioactive fallout. By the time the twenty-three crew members of the Japanese fishing vessel *Lucky Dragon* pulled into port two weeks later, nearly all of its occupants had developed classic symptoms of radiation poisoning. At a press conference on March 31, Strauss reassured the public that the Japanese crewmen would quickly recover and denied that radioactive fallout resulting from nuclear tests posed any danger to American health.

Nearly six months later, Aikichi Kuboyama, one of the Japanese seamen, died. He and one other crewman were found to have traces of strontium-90, a radioactive isotope not normally found in nature, in their bones. Given both strontium-90's long half-life (only half of a sample of the isotope will have decayed in twenty-eight years) and its chemical similarity to calcium and magnesium, the key components of bones, this was a particularly alarming finding. In February 1955, the AEC released a report stating that while it was true that levels of strontium-90 in American soil were increasing, these levels would have to be increased "many thousand times" before they would cause biological harm. Two weeks later, however, panic ensued when a radioactive cloud stemming from a series of small-bore nuclear tests in Nevada set off Geiger counters in Chicago and New York. Once again, the AEC was quick to reassure the public that radioactive fallout was harmless. At a hearing before the congressional Joint Committee on Atomic Energy in April, Strauss reminded his audience that the public's level of exposure to radioactive fallout was miniscule compared to naturally occurring background radiation and medical X-rays. In June, he accused those who predicted biological harm to future generations of acting irresponsibly. The only way to ensure the future of the human race, he argued, was through the continued development and testing of nuclear weapons.

Strauss's continued denials of biological harm caused a furor within the scientific community. Geneticists, in particular, objected to the AEC's claims that radioactivity was harmless at low levels. They had developed expertise in radiobiology on the AEC's dime. By 1960, nearly one in five members of the Genetics Society of America had received AEC funds to study various topics in this field, including the effects of atomic radiation on fruit flies, mice, human tissue culture, and even the survivors of the Hiroshima and Nagasaki bombs.

In 1953, the Atomic Bomb Casualty Commission released a preliminary report in *Science* that suggested that survivors of the bomb were less likely than a control group to give birth to male children, a finding consistent with genetic theories that suggested chromosomal damage from radiation. Over the next five years, a number of scientific organizations—including the National Academy of Sciences, the British Medical Research Council, and an international United Nations scientific committee—issued a series of reports that stated in no uncertain terms that even low levels of radiation caused undesirable genetic mutations in humans.

The public's alarmed response to this scientific consensus emboldened critics of the AEC. Outspoken geneticists such as Alfred Sturtevant, of Caltech, and H. Bentley Glass, of Johns Hopkins University, toured the country warning of the likely increase in birth defects, stillbirths, and human misery that would result from continued nuclear testing. When 1956 Democratic presidential candidate Adlai Stevenson called for an end to nuclear testing, he cited the support of Sturtevant, Glass, and other prominent geneticists. Whereas a decade earlier, Edward Condon had run afoul of federal authorities by advocating international control of atomic energy, critical geneticists continued to receive generous grants from the AEC in the late 1950s. Indeed, Glass lobbed his criticisms from the inside; he was an appointed member of the AEC's Advisory Committee for Biology and Medicine during the period of his most vocal criticisms. As a frequent contributor to the *Bulletin of the Atomic Scientists*, Glass shared the attitude that qualified experts had a duty to offer objective and informed scientific assessments of federal policies. He was careful to identify his opinions as his own, based solely on (what seemed to him) objective criteria rather than a political agenda.

More radical critics called for collective action and questioned whether opposition to U.S. nuclear policies required specialized expertise. In the summer of 1957, Caltech chemist Linus Pauling assembled the signatures of more than 2,000 scientists on a petition calling for an immediate halt to nuclear testing in light of the potential health consequences for current and future human populations. While the list included many high-profile scientists (especially biologists), the majority of the signatories were rank-and-file academic researchers. By the following January, he had added nearly 7,000 more names from around the world, including 37 Nobel laureates. Pauling would eventually be awarded the 1962 Nobel Prize for Peace for his efforts, but not before creating a firestorm in the United States. His critics alleged that such an overwhelming response could not have been organized without outside assistance—presumably from the Communist Party—and in 1960 he was nearly charged with

The Nuclear Test Ban Debate

By the late 1950s, scientific questions about the biological hazards of fallout had erupted into a national debate on the wisdom of a nuclear test ban. Advocates of a test ban argued that mutations caused by exposure to fallout would damage the health of the entire human race for up to ten generations; opponents claimed that exposing the population to the dangers of nuclear annihilation from the Soviets posed a much greater risk. In February 1958, Linus Pauling (*left*) and Edward Teller (*right*) engaged in a heated debate on KQED, a public television station based in San Francisco. Pauling argued that continued testing would result in the births of hundreds of thousands of "defective" children; Teller dismissed these claims as ridiculous. The Pauling-Teller debate created controversy as much for its participants as for its content: who had more authority to question the policies of the Atomic Energy Commission? Was Pauling, a chemist who had been accused of ties to Communism, qualified to speak on national security issues involving biology and physics? Could Teller, a rabid anti-Communist and the leading advocate for the hydrogen bomb, be trusted to offer an objective assessment of the weapon's dangers?

■   From the Ava Helen and Linus Pauling Papers; courtesy of Oregon State University Libraries Special Collections

---

contempt of Congress for refusing to reveal who had helped him gather the names. Yet, this anti-Communistic response was less compelling than it had been at the height of the Red Scare: in the wake of a media outcry defending Pauling's constitutional right to circulate a petition, the Senate's Internal Security Subcommittee quietly dropped its subpoena.

If Pauling's petitions continued to stress the special moral responsibility of scientists to oppose nuclear testing, an alternative movement urged individual citizens to get involved. In 1958 a group of St. Louis scientists, doctors, lawyers, dentists, and housewives founded the Committee for Nuclear Information

(CNI) in response to the conflicting interpretations of fallout issued by the AEC, the NAS, and other scientific institutions. This coalition declared that lay citizens had the right and the responsibility to make decisions about matters of technical policy, including nuclear testing. Experts would share their knowledge in newsletters and public lectures; members of the public would decide. In a brilliant public relations maneuver, the CNI countered the AEC's monopoly on nuclear information by conducting a study of strontium-90 levels in children's baby teeth. Between 1958 and 1963, the Baby Tooth Survey collected over 100,000 teeth from across the nation. The shocking preliminary findings, released in 1961, demonstrated that the levels of strontium-90 in children's teeth had tripled from 1951 to 1955. This grassroots movement to show the effects of fallout on society's most innocent and vulnerable members kept the issue of biological harm in the spotlight long after Strauss and the AEC had hoped to close the debate. By 1962, the CNI had inspired the creation of more than twenty similar nuclear information groups across the country.

Each of these approaches—studies by experts, scientific petitions, and citizen activism—assumed somewhat different roles for scientists in criticizing national defense policy. Glass and his fellow geneticists believed that their specialized expertise qualified them, and them alone, to advise the public on the biological dangers of nuclear testing. Moreover, they defended their close relationships with the AEC, seeing these connections as an opportunity to influence policy. Pauling's assemblage of scientists, on the other hand, suggested that all scientists, as members of an international community that had benefited from the expansion of research funds at the expense of world security, had a moral obligation to protest the irresponsible uses of destructive technologies. CNI's emphasis on citizen activism was based on yet a third assumption—that the consequences of unchecked military growth were so obviously dire that citizens were capable of drawing their own conclusions if only they had access to the proper information.

Given their differences, it would be too much to call these various groups a coalition. Nevertheless, their combined voices had an effect on international nuclear policy. Starting in 1956, the United States and the Soviet Union began a series of protracted negotiations on nuclear test bans and disarmament. Although dreams of a total nuclear test ban faded in the early 1960s with the inability to resolve issues surrounding inspections and verification, Kennedy and Khrushchev agreed to a partial test ban in the summer of 1963. The threat of nuclear annihilation remained, but the terms of the discussion had changed.

Campus Politics

The nuclear test ban debate significantly expanded the range of people deemed qualified to participate in political debates about scientific and technical policy. It did not, however, offer any significant or sustained criticism of the tight bond between defense agencies and civilian institutions. Researchers such as Bentley Glass, for example, saw no inherent conflict in criticizing nuclear testing while accepting AEC funds. So long as the nation was united in its conflict with a common enemy, few citizens—scientists or otherwise—objected to the pursuit of national goals. On America's campuses, few faculty or students understood the extent to which military and AEC funds supported non-weapon laboratories, and fewer still grasped the "dual-use" applications of much research in the social sciences. But when the veil of secrecy and misdirection surrounding the goals of defense-funded research began to unravel, it unraveled quickly. The political consensus that had supported the interlocking civilian and military economy—a consensus that had been two decades in the making—fell apart after only a few short years of focused opposition. By 1970, the campus protest movement had successfully banished the military-industrial complex from its exalted place in the academic research economy. But what would rise in its place?

The first response to large-scale military investment on university campuses was confined primarily to professionals, who tended to interpret it in terms of research ethics and scientific integrity. The response of academic behavioral scientists to international and congressional criticisms of Project Camelot in 1965 was typical. Two years earlier, the Army had begun funding an ambitious program to develop social science models to understand social and political change in developing nations. The Special Operations Research Organization, a nonprofit research institution associated with American University in Washington, D.C., was awarded a three-year, $4- to $6-million contract to oversee the work of psychologists, sociologists, and cultural anthropologists in a number of Latin American nations of interest to military planners. In the summer of 1965, Chilean journalists broke the story that the project was funded by military dollars, not by the NSF, as some Chilean researchers had been told. The subsequent outrage resulted in congressional hearings and a formal protest by the Chilean ambassador but elicited little response on the campus of American University. Behavioral scientists, for their part, interpreted the episode as evidence of the Department of Defense's meddling in projects that should have been undertaken openly by civilian agencies—namely, the Department of State. Few objected to the project's premise that the social sciences could contribute to national security interests. Indeed, at this point, many liberal social scientists hoped that their research might contribute to a more peaceful approach to American foreign policy.

Only two years later protests erupted on the campus of Princeton University—not exactly a hotbed of student radicalism—when the student newspaper reported the existence of a branch of the Institute for Defense Analysis, a Defense Department think tank, on the campus. After students staged a sit-in and faculty members signed a petition, Princeton cut its administrative ties to the institute. Despite this nominal separation, however, the institute remained physically present on the university campus. Similar events unfolded on campuses around the country, as students and faculty at one university after another learned about the classified research taking place in nondescript buildings with innocuous names. On March 4, 1969, the critique expanded to the entire relationship between science and the state, as students and faculty at over thirty universities participated in a day of nonviolent action. This time, scientists themselves—many of whom actually worked on national security projects—joined in the panel discussions, teach-ins, and demonstrations. As the ideological stakes grew, discussion gave way to violent confrontation. In 1970 four students detonated a van filled with more than 2,000 pounds of explosives outside Sterling Hall at the University of Wisconsin in Madison. Though directed at the work of the Army Mathematics Research Center, the explosion missed its target, instead killing one physicist and injuring three others, none of whom was affiliated with the targeted center.

Though not as dramatic as some of the direct action protests elsewhere, the March 4 protests—the brainchild of organizers at MIT—set the terms of debate for national criticisms of what was becoming known as the military-*academic*-industrial complex. A few conservative faculty members at that university, most notably the iconic Charles Draper of the Instrumentation Laboratory, argued that a ban on military funding would restrict scientists' intellectual freedom by preventing them from working on topics of their choosing. But given the by-then widespread opposition to the Vietnam War on the nation's campuses, Draper's position was an anomaly. Increasingly, university administrators and senior faculty accepted the position of student activists that military research had no place on campus. Putting aside questions of morality and ethics, the disruption and physical violence of the student protest movement had upset administrators' calculations of the cost-benefit ratio of accepting military contracts. The question instead became one of whether the universities should cut their ties to federal research contract centers—a strategy known as divestment—or whether they should instead encourage these centers to turn their scientific expertise to socially useful problems—an approach referred to as conversion.

As we saw in chapter 5, some defense contractors had already begun to explore the idea of conversion as an approach to economic diversification. RAND's foray into urban planning, for example, attempted to bring the techniques of systems analysis into the management of modern cities. Closer to home, commentators at MIT pointed to the success of its Fluid Mechanics Laboratory in adding biomedical and environmental projects to a research program previously focused on jet engines and ballistic missile reentry. Beginning in 1966, the laboratory's leadership had actively sought research contracts that might "redress an imbalance" in its research priorities.* By the March 4 protests, almost two-thirds of the lab's budget addressed "socially oriented" problems. Even so, a military presence lingered, not only in the remaining one-third of the budget dedicated to defense problems, but also in the ONR funds that supported much of the "converted" work.

But if conversion seemed a plausible solution for small facilities, it was a utopian fantasy for operations as large as Draper's Instrumentation Laboratory or MIT's Lincoln Laboratory. In 1966, the Fluid Mechanics Laboratory had a budget of $300,000, six professors, and twenty graduate students. In contrast, the 1968–69 budgets of the Instrumentation and Lincoln laboratories were $56 million and $67 million, respectively, with a combined staff numbering in the thousands. Even Ascher Shapiro, one of the leaders of the Fluid Mechanics Lab, urged the university to cut its ties to these defense laboratories. No civilian agency could possibly support either organization at levels remotely close to their current expenditures. More to the point, Shapiro argued, conversion would not stop defense-related research but only push it elsewhere. As he put it, "The only way to bring the military back into reasonable balance is . . . through reduction of budgetary allocations, and through exposure of the symbiotic relationship that exists between the military, certain segments of industry, and certain segments of national and state politics."† University administrators apparently agreed. After a three-year transitional period to sort out the contractual details, in 1973 the Instrumentation Laboratory officially became the Charles Stark Draper Laboratory, Inc., a nonprofit independent research organization. Similar events followed across the country as, one by

---

*Phrase quoted in Matt Wisnioski, "Inside 'the System': Engineers, Scientists, and the Boundaries of Social Protest in the Long 1960s," *History and Technology* 19 (2003): 322.

†Ascher Shapiro, as quoted in Stuart W. Leslie, *The Cold War and American Science: The Military-Industrial-Academic Complex at MIT and Stanford* (New York: Columbia University Press, 1993), p. 239.

one, universities cut their ties to the federal contract research centers that had underwritten their postwar expansion.

MIT's difficulties in eliminating, or even reducing, defense funding from technical laboratories suggested that establishing a role for science independent of national security goals would require two separate processes. First, scientists would need to identify more "socially useful" applications of their talents. The Fluid Mechanics Laboratory's transition to issues of pollution was part of a broader trend, as scientific and engineering journals suddenly proclaimed their interest in environmental problems, urban infrastructure, and public health. Yet, in their willingness to accept non-weapons-related ONR funding, Shapiro and his colleagues expressed tacit approval for the continued role of defense agencies in supporting scientific infrastructure. Truly freeing science from military interests would require a second step, a more fundamental transformation of the political economy of Cold War science.

Given that the expansion of the postwar university assumed that the primary purpose of science was to serve the needs of the state, it is perhaps not surprising that this second, more radical critique found more support off-campus than on. In Congress, the Mansfield Amendment to the 1970 Military Authorization Act stated that DOD funds could only be spent on projects with a "direct or apparent relationship to a specific military function or operation."* Though its definition of a "military function" left some ambiguity and therefore flexibility in the act's implementation, the Mansfield Amendment forced the Pentagon to articulate why military dollars should be used to support civilian projects. The effects of the amendment were immediate and real. In constant dollars, DOD support for basic research at universities declined by more than 50 percent in the first half of the 1970s. Whereas the DOD had supplied more than a third of funding for academic research in 1960, by 1975 that proportion had dropped to less than a tenth. But while this amendment had the desired effect of cutting military budgets for basic research, it did nothing to cut R&D spending on weapons research. Indeed, it had the opposite (and unintended) effect of concentrating military funds at those institutions, such as Johns Hopkins University, that chose not to banish facilities like the Applied Physics Laboratory from their campuses.

Political activists on the New Left, in contrast, questioned whether science should be working for the state, period. Decentralized organizations such as Science for the People, Computer People for Peace, and Science for Vietnam claimed that the institutions of science had betrayed the foundations of par-

*Roger L. Geiger, *Research and Relevant Knowledge: American Research Universities since World War II* (New York: Oxford University Press, 1993), p. 193.

ticipatory democracy. Building on broader critiques of American society that had become popular in the late 1960s and early 1970s, radical scientific activists argued that capitalism, imperialism, sexism, and racism were built into the very structures of scientific knowledge production. "Reform" would be impossible; what was needed instead was an entirely new approach to science that could operate outside of existing political structures. Science for the People, in particular, generated media coverage of its positions by confronting the leaders of American scientific institutions at high-profile professional events. At the 1970 meeting of the American Association for the Advancement of Science, for example, members of the group attempted to "indict" Glenn Seaborg, the chairman of the AEC, for "crimes of science against the people," citing his role in "coordinating and strengthening the dependence of science and universities on war and profit in a unique criminal history of responsibility."* Beyond political theatre, organizations like Science for the People published manifestos, assembled "counter-conferences," organized trips to China and Vietnam, and used their technical skills to help the Black Panther Party evade electronic surveillance. But despite their varied efforts, radical scientific activists found it difficult to reconcile their positions as members of an educated elite with their critiques of scientific knowledge.

Collectively, the student protests, radical critiques, and congressional reforms dismantled the consensus that had ruled university campuses since the end of World War II. No longer would it be acceptable for universities to fuel their expansion with the help of military funds. With the more sweeping radical criticisms offered by organizations like Science for the People, even those scientists who accepted nonmilitary funds increasingly began to ask what sort of ideological strings came attached to federal largesse. And perhaps the biggest change of all was that for the first time since the atomic scientists' movement, scientists felt empowered to offer political criticisms of the relationship of science to national security without repercussions to their careers. Yet, as the scientists would soon find out, defense analysts were no longer so very interested in what they had to say.

Mutual Distrust

Given that protests against the Vietnam War originated on university campuses, campus opposition to the scientific and technical research that undergirded American foreign policy is not surprising. What is perhaps more startling is the speed with which such questioning spread to defense insiders. The

*As recounted in Kelly Moore, *Disrupting Science: Social Movements, American Scientists, and the Politics of the Military, 1945–1975* (Princeton: Princeton University Press, 2008), p. 167.

expansion of U.S. military involvement in Vietnam challenged even those scientists who had previously supported defense work or had served as advisors to the defense establishment. And as their criticisms grew, the distrust between scientists and those they advised became mutual. For the first time in a generation, national security advisors began to ask whether it might be better to make decisions about science and technology without the input of scientists and engineers.

One of the more telling sites for this split was within the Jasons, a secret collective of physicists who spent a portion of their summers providing advice to the Pentagon. Originally created in 1959 under the umbrella of ARPA, Jason scientists offered independent assessments of military technologies and suggested new technologies that military planners might want to pursue. Unlike most defense advisory groups, Jason chose its own members, and the membership itself decided which problems to investigate after receiving top-secret briefings from military leaders. By 1965, Jason's projects had included missile defense, submarine detection and tracking, and schemes to disrupt the earth's magnetic field. Although Jason usually offered solutions to intractable problems, it occasionally used its powers to nix scientifically implausible ideas, such as plans to shoot down incoming missiles with powerful lasers. In short, Jason's members felt not only that their expertise was unique, but also that the military appreciated their independent views.

In 1966 the Jasons decided to address the problem of how to cut off the North Vietnamese supply chain to South Vietnam. In their assessment, Operation Rolling Thunder, an attempt to disable the North Vietnamese through intense bombing, had failed. It would be better, the Jasons argued, to somehow restrict access to the Ho Chi Minh trail, the primary supply route used by the insurgents to move troops and supplies south. This strategy would not only be more effective in dealing with guerrilla resistance in a subsistence-based economy, but would also cost less and reduce loss of life. Secretary of Defense Robert McNamara agreed, and so the physicists set about designing a high-tech "electronic barrier" for the Ho Chi Minh Trail. The electronic barrier was to have multiple components: an array of sensitive audio and seismic sensors, landmines, a remote computer control system, and limited air bombings. But almost immediately after installation of the $800 million system had begun, it became clear that the military—particularly the Air Force—saw the barrier as an aid to, rather than a substitute for, air bombing campaigns. The barrier itself failed to stop the flow of supplies to the south, but the "electronic battlefield" became a central part of modern warfare.

Like the backers of the Franck Report before them, the Jasons learned the hard way that scientists do not necessarily control the things they have created. When summaries of several of the Jason reports appeared in the *New York Times'* publication of *The Pentagon Papers* in 1971, the Jasons became a symbol of all that had gone wrong in the relationship between academic scientists and the military. A 1972 booklet published by the antiwar group Scientists for Social and Political Action, *Science against the People*, recounted the most damning of the Jasons' projects and identified several Jasons by name. Protests followed, including a three-day siege of Columbia University's Pupin Laboratories. Several Jasons resigned, while others simply noted that their function all along had been to advise the military. In the face of personal threats, those Jasons who remained committed to advising the military reiterated their right to advise their own government. They returned to their work with a renewed dedication to secrecy, not so much chastened as disillusioned with the tactics of dissent.

A parallel discussion had been taking place within the President's Science Advisory Committee, whose membership overlapped with the Jasons. Since the early days of the Johnson administration, some of PSAC's younger members had pushed for technological solutions to the escalating violence in Vietnam. More senior members, many of whom had served under Kennedy during discussions of Apollo and the test ban, recognized the limits of scientific and technical expertise in solving political problems. Until 1967, however, PSAC members protected their roles as presidential advisors by publicly supporting the policies of the administration, no matter their personal opinions. That year, tensions over both the electronic barrier and the Johnson administration's decision to develop an antiballistic missile (ABM) system finally brought dissension into the public eye.

PSAC and the Pentagon had been locking horns over the development and deployment of an ABM system since the early Kennedy administration. As its advocates envisioned it, an effective ABM system, backed up with an operational semi-automatic ground environment (SAGE) system, could prevent a nuclear holocaust by intercepting Soviet ICBMs. The majority of PSAC's members opposed ABM on both technical and political grounds. The successful use of an ABM would require split-second decision making, including the ability to distinguish between decoys and actual warheads; an extraordinarily sophisticated launch and navigational system; and a force large enough to counteract an overwhelming first-strike assault. From a strategic perspective, PSAC members argued that the deployment of an ABM system would actually *decrease* nuclear security by destabilizing the logic of mutually assured destruction.

When Secretary of Defense Robert McNamara announced in September 1967 that the Pentagon would shortly deploy Sentinel, a limited ABM system, the news came as a surprise to PSAC members. In January of that year, several PSAC members, including each of the former presidential science advisors, had joined McNamara and Pentagon officials in a meeting to discuss the various ABM options. As they had before in reports and private meetings, PSAC's representatives made their case against deploying ABMs. They left the meeting convinced that their statements had been persuasive; they were not consulted in the intervening months. Feeling betrayed on the one side from the administration, and pressured on the other by their colleagues, who felt that their opposition had not gone far enough, several PSAC consultants criticized the decision in interviews and speeches. McNamara's resignation in November, in large part due to his differences with the Johnson administration, did little to assuage their anger.

That same fall, George Kistiakowsky, a Jason who had served as Eisenhower's science advisor, ended his relationships with the DOD. He did so privately but did not deny the reports when *Science* broke the news, and the *New York Times* repeated it, the following March. The reports stated that Kistiakowsky had become disillusioned with administration policy in Vietnam and preferred to "devote himself to activities that he felt would be more fruitful for reducing the conflict."* Apparently reluctant to sever his ties with the administration completely, he stayed on PSAC as a member-at-large. Nevertheless, this unanticipated decision by such a senior member of the scientific establishment sent shock waves through the defense community and brought renewed attention to Johnson's dismal relationships with his scientists.

President Richard Nixon had promised to address this growing schism between scientists and the federal government during his presidential campaign, but he, too, soon had crossed swords with his scientific advisors over the question of ABMs. When he announced in March 1969 that his administration would pursue a new system, dubbed Safeguard, it was without consulting PSAC. Meanwhile, several members of PSAC had spent the previous week testifying before a Senate subcommittee on the various reasons why they opposed the deployment of ABM systems. This embarrassing situation, combined with subsequent public disagreements with PSAC on environmental and economic issues, suggested to Nixon that scientific advisors created more political problems than they solved.

Upon his reelection in 1972, Nixon promptly disbanded PSAC. By that

*Daniel Greenberg, "Kistiakowsky Cuts Defense Department Ties over Vietnam," *Science* 159 (March 1, 1968): 958.

point, the scientists were not sorry to leave. Yet Nixon's decision did not mean that issues of science and technology were no longer central to national security. Nixon continued to solicit opinions from technical advisors on issues related to national defense—but now these opinions came from the Pentagon, which in turn leaned heavily on its contractors for advice. For analysis of domestic issues, he created the Office of Science and Technology Policy. And for general issues related to the health of the American scientific research enterprise, he would turn to the director of the NSF. The brief era of independent presidential science advising, created with such fanfare after *Sputnik*, lasted a mere fifteen years.

The Vietnam-era controversies surrounding scientists' work as defense contractors and advisors highlighted not only the moral ambiguity of defense research, but also the broader problem of scientists' relationship to the state. Is it wrong for scientists to contribute to work that might do harm? What if the primary purpose of that work is defensive rather than offensive? Might it be acceptable for academic scientists to design, say, a missile defense system but not a weapons-delivery system? Or is the morality of defense work defined by the justness of the fight, as when individuals distinguished between scientists' contributions to World War II and Vietnam? And if scientific prestige is itself a weapon in a global battle for national supremacy, under what circumstances might a scientist accept nonmilitary federal funds for basic research?

These were (and are) questions with no simple answers. What was clear to everyone involved, however, was that the notion of the proper relationship between civilian and military science had shifted. After the passage of the 1970 Military Authorization Act with its Mansfield Amendment, military dollars could no longer easily support nonmilitary projects; as a result of campus protests, classified research moved (mostly) off-campus. But if these events had the effect of "cleansing" academic science of military taint, they did little to discourage the military's love affair with high-technology weapons. The Institute for Defense Analysis and the Charles Stark Draper Laboratory, Inc., still existed; the difference was that they now operated without university oversight. These changes mirrored a broader separation between military and civilian life taking place within American culture, a trend that culminated in the end of the military draft in 1973.

But there was yet another shift afoot. By the late 1960s both the United States and the Soviet Union had enough bombs and missiles to destroy one another's populations many times over. National security advisors increasingly spent their time analyzing risk instead of fantasizing about new weapons; political advisors in both nations urged their leaders to invest in social programs that had been neglected during the arms buildup. After years of acceding to the Pentagon's requests, even Congress began to question the relationship between scientific research and national security. In the mid-1960s, at the request of Congress, the Pentagon undertook a PPBS-style assessment of its investment in basic and applied research. To scientists' chagrin, "basic research" was found to have accounted for less than 0.5 percent of advances in weapons research. Applied research fared somewhat better, but the real value, it was clear, was in development, not research. The report, called Project Hindsight, stated in no uncertain terms that it was unable "to demonstrate value for recent undirected science."*

In retrospect, Project Hindsight stands out as a turning point in the increasingly difficult relationship between scientific researchers and the military. If basic research failed to produce weapons breakthroughs, why should the Pentagon put up with scientists who questioned their strategies, whether in Vietnam or Washington? Hindsight's emphasis on value also proved telling, as congressional representatives from both sides of the aisle increasingly demanded that scientists demonstrate the economic value of their research in the 1970s and 1980s. Both of these issues—the rejection of scientific expertise and market value—would take center stage in the last decade of the Cold War.

*Geiger, *Research and Relevant Knowledge*, p. 191.

8 Cold War Redux

The United States and the Soviet Union probably came closer to nuclear war in the early 1980s than they had at any time since the Cuban missile crisis. The last decade of the Cold War—sometimes called the Second Cold War—witnessed plenty of tough talk, with defense budgets to match. As they had throughout the postwar period, both nations invested heavily in pie-in-the-sky technologies as the key to national defense. Once again, scientists and engineers at federal, private, and university laboratories turned their attention to developing advanced nuclear weapons systems. And as before, large parts of this research took place behind closed doors, inaccessible to all but those with the highest security clearances. The military-industrial complex that Eisenhower had warned about was back, and with a vengeance.

Reagan's Cold War, however, was not Eisenhower's. American political leaders in the 1980s embraced a free market ideology that placed a priority on limiting the size of the federal government. Right-wing idealogues questioned the government's role (aside from the essential functions of defense) in American life. It was no longer obvious to them, or to their allies in Congress, that the federal government should provide lavish funding for basic research in science and technology. In an era of globalization, policymakers encouraged the development of scientific fields that could compete in the global marketplace as well as in the marketplace of ideas.

These twin concerns—national defense and privatization—shaped the course of American policy for science and technology in the waning days of the Cold War. This final chapter explores the general role of science and technology in addressing these challenges before turning to specific examples of their embodiments in science and technology policy.

Dual Threats

The 1970s and 1980s were a period of tremendous instability for international politics and the global economy. Within the United States, the ascendant conservative political movement called for a renewed commitment to anti-

Communism and, with it, a reduction in the size and function of the federal government. At the same time, economic growth in Japan, Germany, and a unified Western Europe presented new challenges to American dominance of global markets. Visions of science and technology were once again front and center in shaping the American response to the dual threats of Communism and economic competitiveness.

The 1970s started with a thaw in the superpowers' relationship, a process that the Americans referred to as détente. In February 1972, President Richard Nixon went to China, where he met with Chairman Mao; three months later, Nixon signed an ABM treaty with Soviet premier Leonid Brezhnev. The "Basic Principles" endorsed by both leaders stated that each nation would avoid "efforts to obtain unilateral advantage at the expense of the other," meaning that neither would attempt to shift the global balance of power. Détente was, in effect, stalemate. Like most compromises, détente satisfied no one. Advocates of disarmament decried the missed opportunity to reduce nuclear arsenals; hawks viewed any agreement with the Soviet Union as capitulation.

By the time newly elected President Jimmy Carter turned in 1977 to implementing his campaign promise to reduce, or possibly even eliminate, nuclear weapons, détente had already begun to fall victim to volatile politics in the Middle East and Africa. The United States' involvement in the 1973 Arab-Israeli War and the Soviet Union's participation in a number of African civil wars, most importantly in Angola and Ethiopia, suggested that neither nation was quite willing to give up its role in dominating global politics. The Soviet Union's invasion of Afghanistan in December 1979 announced the formal end of détente to any observers who might have missed its more gradual decline.

The renewal of Cold War tensions was accompanied by dramatic increases in defense spending. Under President Ronald Reagan, a conservative who described the Soviet Union as the "evil empire," defense spending as a fraction of GDP rose by nearly a third, from a postwar low of 4.7 percent in 1978 to 6.2 percent in 1987. This increase mirrored a shift in the source of federal R&D monies. In 1979, the DOD supplied well under half (44 percent) of the federal investment in science and technology; by 1987, at the peak of the Reagan-era defense buildup, the DOD was once again bankrolling nearly two-thirds (63 percent) of scientific R&D. This number is even more startling in the longer context of the Cold War: the proportion of federal R&D coming from defense in 1987 was higher than it had been at any point since 1962. The Great Society's promise to use federally funded scientific research to conquer disease, pollution, and poverty had gone unfulfilled.

As a consequence of both the Mansfield Amendment and Project Hindsight, DOD dollars were increasingly concentrated at industrial contractors and nonprofit research centers held at arm's length from universities. For example, the Charles Stark Draper Instrumentation Laboratory, Inc., now operating as an independent entity, consistently ranked first among nonprofit federal R&D contractors throughout the 1980s. Meanwhile, fewer American corporations depended on federal contracts for the bulk of their profits. Having suffered from their reliance on single-source funders during the defense cuts of the late 1960s and early 1970s, such leading defense contractors as Westinghouse, GE, Honeywell, and Raytheon either diversified their products and services or left the defense field altogether. The uptick in defense spending certainly strengthened the remaining contractors' bottom line, but these contracts represented a much smaller part of a larger portfolio. Throughout the 1980s, for example, only five of the top fifteen defense contractors obtained more than 60 percent of their revenue from contracts with NASA or the DOD. With the exception of this core of defense contractors (Grumman, General Dynamics, Northrop, Martin Marietta, and McDonnell Douglas), an increase in military spending no longer translated directly into corporate profits.

For many university administrators and corporate research managers, the military buildup was a distraction from structural changes taking place within the American economy. The American economy had all but collapsed in the 1970s, with real wages falling and productivity flat. With instability in the Middle East, oil prices soared; inflation soon followed. The target interest rate of the U.S. Federal Reserve for bank-to-bank lending, a benchmark of the overall economy, climbed steadily from its historical postwar average of below 5 percent to a peak of nearly 20 percent in 1981, making capital investments prohibitively expensive. Economists and business leaders expressed particular concern over the emergence of a permanent trade deficit: beginning in 1976, the United States has consistently imported more goods and services than it has exported. Japan, in contrast, experienced phenomenal economic growth. For the first time since the Great Depression, the challenge of economic competitiveness ranked equally with the fight against Communism in the minds of both the public and those who established federal science policy.

An increasingly powerful conservative political movement placed the blame for this economic downturn on the growth of the federal government. Neoliberals like the University of Chicago economist Milton Friedman blamed excessive federal regulation and taxation for posing disincentives to corporate innovation. First under Carter and then more aggressively under Reagan, health, safety, and environmental regulations became subject to the same sorts of cost-benefit analyses previously reserved for federal spending. Conservative economists, moreover, argued that dramatic changes would be needed in science and technology policy to encourage inventors to bring new products to market. Even before Reagan's election in 1980, conservative economists were recommending a shift in the role of the federal government from underwriting basic research that might someday produce breakthrough technologies (the *Science: The Endless Frontier* model) to encouraging the development of commercial applications from existing scientific knowledge (a concept known as technology transfer).

These recommendations soon took the form of concrete policy actions that transformed the research economy in the 1980s. The two most important of these were the 1980 Patent and Trademark Amendment Act (also known as the Bayh-Dole Act after its sponsors in the Senate) and the 1981 Economic Recovery Act. The Bayh-Dole Act allowed universities, small businesses, and nonprofits to file patents on federally funded research, a privilege rather than a right under previous law. Similarly, the Economic Recovery Act provided tax credits for businesses that invested in scientific and technical R&D. Universities began courting industrial patrons on a level not seen since before World War II, with many campuses building research parks dedicated specifically to conducting investigations on behalf of corporate sponsors. Even the NSF, the last redoubt of basic research, increased its support for applied research and encouraged the creation of public-private cooperative research centers. By the end of the decade, at least 10 percent of the NSF's budget was dedicated to engineering, with researchers in all fields increasingly asked to identify the practical applications that might result from their work.

Critics soon charged that corporate funds were distorting research agendas, warping graduate education, and inhibiting the free exchange of information. Though few commented on it at the time, their complaints were eerily reminiscent of those from critics who had lamented the influence of military dollars on campuses in the 1950s. Yet beneath these cosmetic similarities, there lurked a crucial difference: in the early Cold War, the benefits of this system accrued to the public, in the form of supplying the needs of the federal government; in the 1980s, they accrued to private industry, in the form of corporate profits. As a means of illustrating both continuity and change, the next two sections take a closer look at scientific entrepreneurship and military R&D in the 1970s and 1980s.

The New Academic Entrepreneurship: Biotechnology

Physics may have captured the spotlight for most of the Cold War, but some of the era's most dramatic advances came from the field of biology. James Watson and Francis Crick's 1953 discovery of the structure of DNA was only the most famous of a series of discoveries that had, collectively, transformed biology from a descriptive science to a powerhouse of experimental method. When researchers at Stanford University and the University of California–San Francisco announced in 1973 that they had successfully inserted a specific sequence of DNA into a target organism, the stage was set for a biological gold rush. Within a few short years, biotechnology became the darling of the stock market; academic biology would be forever transformed.

As pioneered by UCSF's Herbert Boyer, Stanford's Stanley Cohen, and their colleagues, recombinant DNA technology functions by tricking bacteria into treating a section of spliced DNA containing a gene sequence from a higher organism as their own. The technique's commercial promise rested on the prolific reproductive capacity of *Escherichia coli*, the most common target organism. Assuming that researchers could learn how to control the system, *E. coli* could be used as a sort of bacterial factory to produce biological substances on demand. Instead of treating diabetes with insulin derived from slaughtered pigs, for example, it might be possible to produce commercial quantities of human insulin using recombinant DNA. Nobel laureates and science reporters alike predicted astonishing uses and dramatic applications, from manufacturing chemicals and human "spare parts" to creating novel biological weapons. Not coincidentally, commentators took to calling the technique "genetic engineering."

Still, commercial applications for genetically engineered organisms were a ways off in 1973 when Stanford and the University of California applied for a patent to cover the technique. The research had been funded with a grant from the NIH, which had recently implemented intellectual property agreements with sixty-five universities (including Stanford and California) in an attempt to encourage the commercial development of biomedical technologies. Stanford, meanwhile, was at the forefront of the movement to use revenues from patents to offset losses in defense funding. In 1970, it had established an Office of Technology Licensing to encourage faculty members to disclose inventions, file patents, and market licenses. Though the envy of university administrators, Stanford's aggressive patent policy was not without its critics, particularly within the field of molecular biology. Aside from the question of federal funding, scientists complained that Cohen and Boyer's achievement was not enough of an advance over existing techniques to merit the granting of a patent.

After a six-year legal battle, the U.S. Patent Office issued Patent No. 4,237,224 to Stanford in 1980. Within two weeks of the granting of the patent, Stanford had licensed the technology to seventy-two companies for annual payments of $10,000, plus royalty fees of up to 1 percent of sales on products using recombinant DNA technology; it soon had the highest patent income of any university in the country. By 1984, the income from this single patent exceeded more than $750,000, with lifetime earnings expected to yield somewhere between $250 and $750 million. The earnings were so high because the scope of the patent was so vast: theoretically, anyone using recombinant techniques without a license, regardless of the specific biological substances involved, could be challenged with patent infringement.

Even during the period of uncertainty for the Boyer-Cohen patent, both corporate and academic entrepreneurs had begun to reconsider the role of intellectual property in an investment portfolio. Before the 1970s, most universities refrained from seeking patents on discoveries related to public health, under the theory that universities had a public service obligation to share the results of their research with the public. A harbinger of change was an agreement in 1974 between Harvard Medical School and the Monsanto Corporation. For a fee of $23 million, spread over twelve years, Monsanto would receive exclusive license to patents generated by antitumor research funded by the corporation at Harvard. Although this was hardly the first time that a major corporation had partnered with university researchers, the scale of the grant, and its focus on what the scientists involved insisted was "basic" research, was unprecedented.

Monsanto's investment in Harvard was based on the idea that a small upfront investment in scientific research might eventually yield tremendous long-term income in the form of patents. In contrast to the postwar model, the idea was not that basic research would inevitably lead to useful applications that could then be bought and sold; rather, it was that basic research itself could become a commodity through the patent process. Knowledge could be transformed into "intellectual property."

Over the 1970s and 1980s, the commercialization of biotechnology proceeded along several main paths, all starting with this notion of intellectual property. The most familiar route followed the model of the Harvard-Monsanto partnership: multinational pharmaceutical companies partnered with academic researchers and small firms, both as a means to secure patent portfolios and to keep an eye on new developments that might affect their existing product lines. The other routes to commercialization were more novel.

Beginning in the early 1970s, a number of academic researchers, including Herbert Boyer, formed start-up companies while maintaining their university affiliations. While some of these researchers retained ownership in hopes of bringing a product to market, a more typical strategy was to partner with investors from venture capital, a new form of high-risk, high-return investment then popular in California's Silicon Valley. Venture capital made sense only in the context of a science and technology policy that prioritized economic growth; tax rates on capital gains fell from 48 percent to 28 percent from 1968 to 1978. Because the business model relied on the promise of future investors recognizing the market value of a company's research portfolio—including, of course, its patents—a start-up powered by venture capital might be considered successful even if it never brought a product to market. By 1980, venture capital and multinational corporations had invested more than $42 million in a handful of genetic engineering firms, most of which were founded by academic scientists or included them on their board.

Boyer's company, Genentech, pioneered this approach. When Boyer filed for incorporation, he did so with the assistance of Robert Swanson, an unemployed venture capitalist formerly associated with the firm of Kleiner & Perkins. Swanson had already received a promise of seed money in the form of $100,000 from his former employers, for which the company would receive 20,000 shares of Genentech stock. As founders, both Boyer and Swanson received an initial 25,000 shares. In June 1978, Genentech signed a contract with Eli Lilly to the tune of $50,000 a month for the biotech firm's attempts to synthesize human insulin. Lilly, it should be noted, was simultaneously subsidizing similar research at one of Genentech's competitors, all in hopes of protecting its existing market share in treating diabetes. Genentech's stock market debut in October 1980 defied all expectations, with shares jumping from $35 to $80 within minutes of the opening bell. By the time the market finally closed, each of the 1.1 million shares was worth $71, and, at least on paper, Genentech was worth $532 million. For a company without products or revenue, this was a remarkable achievement.

Eventually, Genentech did manage to bring products to market, as did its competitors. The stock market bubble for biotech also eventually burst, as most bubbles do. Even so, the commercialization of biotech set a pattern for a series of research-driven industries in the 1980s and 1990s. University faculty members in science and engineering departments have become well versed in the ways of patents, articles of incorporation, and exit strategies as investors have jumped on the technology bandwagon. And while some faculty members are deeply uncomfortable with what they consider to be the commercialization of the academy, quite a few others have gotten rich. Those fields with little conceivable relationship to the market, in contrast, have suffered budget cuts and attacks on the need for their existence. With the end of the Cold War imperative, commercial value has come to supplant national defense as the defining characteristic of academic science.

The Iron Triangle: Star Wars

The sociopolitical and economic changes of the 1980s ensured that the military-industrial complex that emerged in that decade differed significantly from the one of the 1950s and 1960s. For better and for worse, the defense buildup did not yield a blank check for basic research or university infrastructure. Nor did it underwrite research scientists' salaries at major corporations. Instead, the new approach to defense R&D concentrated contracts for specific projects— so-called programmatic research—in the hands of a few specialized defense contractors and federal contract research laboratories. On the one hand, fewer researchers had to make their views, scientific or otherwise, conform to military expectations; on the other, the pool of qualified experts willing or able to criticize expensive, half-baked military plans shrank considerably. The most notorious result was President Reagan's Strategic Defense Initiative (SDI), a plan known to its critics as "Star Wars."

As a defense hawk and an outspoken anti-Communist, Reagan found the logic of mutually assured destruction difficult to accept. The 1972 agreement to limit ABMs had given official sanction to the idea that the best defense against a nuclear holocaust was no defense at all. Since neither the United States nor the Soviet Union would be able to stop an incoming nuclear attack, the recipient would have no choice but to respond with an equally destructive counterattack. Since by now both superpowers had more than enough nuclear weapons to destroy the world many times over, launching a nuclear attack was equivalent to suicide. Therefore, so the theory went, having the ability to intercept an incoming ICBM would actually increase instability by removing the guarantee of nuclear annihilation.

Mutually assured destruction was, and is, a strategy that requires its adherents to recognize that nuclear weapons are essentially unusable. Thoughtful critics also pointed out that the strategy left no protection against accidental launches or rogue leaders who did not grasp the concept of deterrence. For Reagan, the idea that the United States had absolutely no recourse in the event of a nuclear attack was inconceivable. Critics on the left had, of course, been making the same argument for years, but their solution stressed disarmament rather than missile defense. Having (at this point) rejected disarmament out of hand, Reagan instead resolved early in his administration to commit the United States to a defensive strategy.

Previous plans for nuclear defense, such as Sentinel and Safeguard, had centered on ground-launched missiles, but these plans had obvious limitations. As a condition of the 1972 ABM Treaty, the location of these sites needed to be made public for international inspection, but inspection put the sites at risk of nuclear attack. Moreover, the treaty limited both the United States and the Soviet Union to two ABM sites: one to protect each nation's capital, and one to protect its ICBMs. The DOD's scientific advisors also doubted that it would be possible to design ABMs able to overcome the threat of decoys and multiple reentry warheads. ARPA and the remaining Jasons had been exploring possible alternatives throughout the 1970s but had come up with little beyond lasers and particle beams, which most scientists regarded as science fiction.

Few of these doubts reached Reagan. Like Carter and Ford before him, Reagan operated without a PSAC; George Keyworth, the president's science advisor, informed the scientific community that the president was not interested in hearing the opinions of those "who do not share the Administration's view."* Instead of receiving briefs on the strategic and technical limitations of nuclear defense, Reagan heard from such defense advocates as former Lawrence Livermore National Laboratory director Edward Teller, who assured him that space-based defense systems could protect the nation from nuclear attack. Indeed, Teller began lobbying members of Congress on the possibility of a space-based, nuclear-powered X-ray laser just days after Reagan's inauguration. Meanwhile, Wyoming Senator Malcolm Wallop had shared his own vision for a space-based chemical laser defense system with the president. Members of Congress debated the merits of the two laser-based defense systems throughout 1982, despite the skepticism of nearly every scientist familiar with the plans.

The so-called laser lobby depended in part on a new form of private political activism. In 1981, retired Lieutenant General Daniel O. Graham, a former director of the Defense Intelligence Agency, convinced a group of well-heeled conservative activists that the United States was in grave danger of losing its strategic advantage to the Soviets. Under the sponsorship of the Heritage Foundation, a right-wing think tank, Graham's High Frontier Panel articulated a vision of missile defense that linked space technology, exploration of the skies, and free enterprise. Although lobbying was hardly novel in 1981, the idea that a group of private activists could influence a presidential decision by

*George Keyworth as quoted in Gregg Herken, *Cardinal Choices: Presidential Science Advising from the Atomic Bomb to SDI*, rev. and expanded ed. (Stanford: Stanford University Press, 2000), p. 202.

The Strategic Defense Initiative

President Ronald Reagan's Strategic Defense Initiative embraced high-tech military devices roundly dismissed by leading members of the scientific establishment. This artist's depiction, commissioned by the Department of Defense around the time of the program's announcement in 1983, shows land-based lasers targeting enemy satellites via giant space-based mirrors. Other schemes included using space-based lasers to mount a defense against incoming ICBMs. SDI's proponents viewed the program as a critical investment in American defense and economic development; its critics derided it as science fiction.

■ Courtesy of the U.S. Department of Defense

bankrolling its own foreign policy study was relatively new. Although differing in the details, the report that the group presented to Reagan in January 1982 contained most of the concepts that would eventually find their way into SDI.

The High Frontier Panel's report linked military dominance of the skies to future economic development. The Soviet Union was stockpiling enough ICBMs to overwhelm the entire American arsenal, it claimed, thereby negating the concept of deterrence. Meanwhile, political support for building an offensive missile system that could withstand a Soviet attack was on the decline, as the "nuclear freeze" movement gained momentum with public rallies and support from religious leaders. A space-based system that could cripple Soviet ICBMs in the launch phase could address both problems by improving the accuracy of American ABMs (by allowing the American missiles to disable the Soviet missiles before they split into multiple warheads) and eliminating local opposition to nuclear weapons (by removing the problem of siting missile silos on Earth). Moreover, through the process of building a space-based defense system, the government would be establishing an infrastructure for future commercial development. In much the same way that the British Empire was made possible by the Royal Navy's control of the seas, the nation that controlled space would hold a significant competitive advantage over its economic competitors.

Although Reagan would not announce his plans for SDI for an additional year, he apparently found these arguments compelling. As a former governor of California, home to some of the largest aerospace contractors, Reagan was particularly taken with the idea of commercialization of space. The space shuttle had originally been envisioned as a low-cost carrier for public as well as private satellites, but by 1982 it had already become clear that it would fall short of commercial expectations. Aside from ongoing maintenance issues, the shuttle's most important limitation was that its low Earth orbit made it incapable of launching valuable geosynchronous satellites to their orbits 22,300 miles up. The French aerospace firm Arianespace had capitalized on this market opportunity with its Ariane rocket, which began offering commercial launch services for high-orbit satellites in 1981. On Independence Day 1982, Reagan announced a major shift in federal space policy that emphasized the "domestic commercial exploitation of space capabilities, technology, and systems for national economic benefit . . . consistent with national security concerns, treaties, and international agreements."* The fate of the aeronautics industry, past, present, and future, was therefore very much on Reagan's mind as he considered the possibilities for nuclear defense.

Despite Reagan's well-known interest in both space and nuclear defense, his announcement of a space-based Strategic Defense Initiative in March 1983 came as a surprise to almost everyone, including his closest advisors. The sincerity of Reagan's desire to protect the nation from nuclear attack via space-based lasers does not change the fact that space-based lasers did not, at that time, exist. The response from the academic scientific community was swift and unforgiving. Virtually all of the leading figures of 1950s- and 1960s-era science defense policy—including Jerome Wiesner, Lee DuBridge, and I. I. Rabi— condemned SDI as a costly boondoggle that would destabilize international relations. Editorials in *Science*, the *Bulletin of Atomic Scientists*, and even the *New York Times* criticized Star Wars as bad science and bad foreign policy. Star

---

*Ronald Reagan as quoted in Joan Lisa Bromberg, *NASA and the Space Industry* (Baltimore: Johns Hopkins University Press, 1999), p. 121.

Wars nevertheless became official government policy in 1984 with the creation of the Strategic Defense Initiative Organization, with funding for the first year alone set at $1.4 billion.

The course of the Star Wars debate shows how the relationship between science, technology, and national security had fractured in twenty years' time. Where once the defense establishment fretted about keeping academic physicists on tap in the event of a national emergency, the establishment now turned to its own experts in sympathetic nonprofit think tanks, defense contracting companies, and national laboratories. Of the $7.6 billion spent on SDI from March 1983 to June 1986, 95 percent went to private contractors and federal laboratories, with only 2.3 percent going to universities. In addition, these university contracts were spread thin, with an average contract value of $1 million. Defense contractors had meanwhile gained an outsized role in the political process, in part through a 1974 federal election law that granted federal contractors the right to contribute to political campaigns. Collectively, political action committees associated with the top twenty SDI contractors contributed almost $4 million to candidates running for federal office in 1984. Decisions on defense R&D policy were increasingly controlled by what came to be known as the "Iron Triangle" of defense contractors, defense agencies, and Congress, with outsiders largely shut out of the process.

Once under way, Star Wars proved difficult to rein in, despite a dismal experimental record, nearly continuous criticism from the scientific community, and the eventual collapse of the Soviet Union. Presidents George H. W. Bush and Bill Clinton scaled the program back to a regional, primarily ground-based antiballistic missile system, but under President George W. Bush, high-powered lasers once again entered the (theoretical) armamentarium. Since 1985, the United States has spent well over $100 billion on various high-technology nuclear defense programs, some more plausible than others. While having some sort of means to intercept incoming missiles is an utterly reasonable, even essential, part of national defense in a post–Cold War world, the entrenchment of the Iron Triangle has made a rational discussion of missile defense impossible. On the rare occasions when policymakers have reconsidered the structure and function of missile defense, research scientists—whose opposition to the program is well known—have not generally been consulted.

The End of an Era

Defying all odds, the Cold War between the United States and the Soviet Union ended peacefully in the late 1980s. The normalization in relations occurred in spite of, rather than because of, the nuclear weapons that had for so long symbolized the tensions between the two superpowers. A full discussion of the reasons for the decline and eventual collapse of the Soviet Union is beyond the scope of this book, but most historians now agree that the fall of the regime had as much to do with internal disarray as it did with pressure from the United States. A global commitment to the concept of human rights and national sovereignty had made the position of the Soviet Union vis-à-vis its satellites and its citizens increasingly untenable. Without the personal commitment of Soviet leader Mikhail Gorbachev to *glasnost* (openness) and *perestroika* (restructuring), the end result might have been very different, but as it was, the Soviet Union went out with a whimper, not a bang.

Which is not to say that weapons were not part of the equation. From all accounts, it appears that the Soviet Union was nearly bankrupt by the mid-1980s, having spent vast sums of money building a nuclear arsenal. Besides its enormous stockpile of ICBMs, the Soviet Union had invested heavily in expensive antisatellite weapons. Whatever the American scientific community's assessment of SDI, it seems to have genuinely spooked the Soviets into spending money that they did not have on defense. Exact figures on Soviet defense spending are hard to come by, but American experts at the time believed that, as a function of gross national product, the Soviets were spending about two-and-a-half times as much as the Americans. Perhaps because the funds to counter SDI were not available, the Soviets frequently returned to the arms control bargaining table in the 1980s. Even so, it would be an error of hindsight to say, as some champions of Reagan have, that the United States won the Cold War by forcing the Soviet Union to spend its way into disarray. By the mid-1980s, SDI was only one of many problems facing the Soviet Union.

The United States, meanwhile, entered a brief period of economic boom, fueled by investments and advances in a series of high-tech industries, most notably biotechnology and information technology. In retrospect, however, it has become clear that the economic gains of the 1980s and 1990s were spread unevenly across the population. Corporate profits, driven by changes in the tax code, masked the precipitous decline of American manufacturing. U.S. factory owners moved production to Asia and Latin America to take advantage of cheap labor and infrastructure—infrastructure that had often been funded, at least in part, by international development projects. Nor would the globalization of capital have been possible without new communications technologies developed by American defense laboratories and relayed by American and European satellites. The new neoliberal consensus, meanwhile, had succeeded in shifting American colleges and universities to private, rather than public support, most notably in the forms of dramatic tuition increases and corporate partnerships. With public support for higher education on the wane, academic scientists increasingly felt the pressure to focus their research on areas ripe for outside investment. Science, which for so long had ridden the coattails of postwar American power, suddenly had to pay its own way.

Epilogue

The abrupt end of the Cold War presents an ambiguous lesson for those trying to understand the relationship of science to the state. Throughout most of
the postwar period, American policymakers assumed that the nation's future
depended on science and technology, but they did so for very different reasons.
Political conservatives tended to see the role of science as primarily for national
security, or barring that, for national economic competitiveness. From this
perspective, the purpose of cultivating a rich and vibrant scientific community
was to ensure that the United States had access to technologies of domination, whether that preeminence took the form of a weapons stockpile, a space
transportation network, or intellectual property.

An alternative interpretation, usually held by political liberals, looked to
science and technology as a public good that was, by definition, worthy of
public support. A close corollary was the idea that widespread access to the
products of science and technology would serve to demonstrate the superiority
of the American way of life. At the height of the Cold War, this interpretation
took on an additional meaning, as a self-governing scientific community came
to be seen as a stand-in for American society. In its openness, its supposed
egalitarianism, its internationalism, its can-do attitude, the scientific community represented the best that American society had to offer to its own citizens
as well as to those of the world.

Both of these understandings of the proper role of science and technology
depended on the existence of an ideological foil. Overnight, the collapse of
Communism stripped the American scientific community of much of the justification for its existence. Without a rival superpower, wars might be fought
with more mundane technologies that did not require the input of the world's
leading scientists. And since the United States had won the Cold War, the
superiority of American life was taken as a given, rather than something that
had to be demonstrated in the court of international opinion.

The scientific community has yet to find its footing on this new terrain, a topography made more slippery by the concomitant rise of a global market-place. With even the People's Republic of China adopting a market economy, neoliberal economists have claimed victory for privatization. Putting aside the question of how commercialization shapes the practice of science in fields that are amenable to it, what will become of those areas of science, like astronomy or high-energy physics, that lack reasonable prospects for market exploitation? Since the terrorist attacks of September 2001 and the subsequent wars in Iraq and Afghanistan, some university researchers have once again turned to the defense establishment, but, as with SDI, the greatest portion of recent military R&D funds have flowed directly into the coffers of defense contractors rather than into university infrastructure.

The end of the Cold War has also undermined, in America at least, the for-merly unassailable consensus that science should be an independent, objective enterprise. From the establishment of the AEC as a civilian institution to the notion, enshrined in the NSF, that scientists were the best judges of what sorts of research the nation might fund, Cold War–era science policy gave lip service to the idea that scientific knowledge existed on a plane apart from politics. American policymakers from the end of World War II through the start of the war in Vietnam assumed that scientists had something special to contribute to discussions of defense technologies. Even if defense analysts and Pentagon offi-cials disagreed with scientists on matters of policy, most believed that weapons should not be developed without the tacit endorsement of technical experts. If Nixon's disbanding of PSAC represented the first step away from this practice, Reagan's announcement of SDI represented a complete repudiation of advice from the scientific community. To conservative commentators, the fact that SDI triumphed in what they had come to see as the "marketplace of ideas" trumped SDI's tenuous relationship to reality. The idea that scientific reality can be custom-built to fit a specific point of view lives on in industry-funded scientific research, such as that which has continued to deny the existence of climate change in the face of overwhelming evidence to the contrary.

This is not to say, of course, that the federal government has entirely aban-doned either the institutions or the spirit of science since the end of the Cold War. A decades-long arms race left the United States with an archipelago of research sites whose scientists, contractors, and congressmen formed a power-ful political constituency. The national laboratories, in particular, have proved resilient to cuts in federal R&D spending, continually finding new projects to justify their existence. In 1977, President Jimmy Carter created the Depart-ment of Energy as a successor to the Atomic Energy Commission, which had just three years earlier been separated into regulatory and production branches. The DOE inherited not only the AEC's physics laboratories but also its super-computers and research programs in genetics and radiobiology.

While several of the national laboratories have turned toward research man-dates in developing new forms of renewable energy, the highest-profile conver-sion took place at Los Alamos, where scientists embarked on an ambitious plan to catalog and sequence the human genome. The resulting Human Genome Project not only established a new institutional partnership between the DOE and the NIH but also created a new federal constituency for biological research within a traditionally defense-oriented institution. And as might be expected in the 1980s' climate of commercialization, the Human Genome Project paid particular attention to encouraging technology transfer and entrepreneurship. A recent report publicized by a biotech industry trade group claims that the $3 billion the federal government invested in the Human Genome Project cre-ated nearly $800 billion in economic growth and created more than 300,000 new jobs.* The Great Society may be long gone, but the dream of science and technology as an economic engine lives on.

Other Cold War habits have similarly reappeared in American public pol-icy in recent years. At the 2011 Annual Meeting of the American Association for the Advancement of Science, John Holdren, President Barack Obama's advisor for science and technology, described the president's embrace of sci-entific R&D to a standing-room-only crowd. Holdren recounted the many ways in which President Obama had attempted to restore prestige to American science, listing the number of Nobel Laureates, National Academy members, and former AAAS presidents in positions of influence in Obama's govern-ment. But if Holdren's praise of the administration's generous research budgets and respect for scientific expertise might have momentarily caused listeners to check the date on their calendars, a quick glance at the Republican presiden-tial primary debates in the fall and winter of 2011 to 2012 would have swiftly returned them to the present. One after another, the Republican presidential candidates announced their disdain for scientific authority. (The only outlier among these candidates, former Utah governor Jon Huntsman, did little to increase his already tepid support in the primary polls with his announcement on the social media platform Twitter: "To be clear. I believe in evolution and

*Roy Zwalen, "Battelle Report: $3.8 Billion Investment in Human Genome Project Drove $796 Billion in Economic Impact Creating 310,000 Jobs and Launching the Genomic Revolu-tion," *Patently Biotech* (blog), May 5, 2011, *BIOtechNow*, www.biotech-now.org/public-policy/patently-biotech/2011/05.

trust scientists on global warming. Call me crazy."*) Holdren's comments to the largely moderate-to-liberal academic members of AAAS were intended as reassurance to a reliable political constituency, not as an announcement of a dramatic shift in federal research policy. Support for science had become a partisan, rather than a national, rallying cry.

Even technologically driven schemes for global development have reappeared in recent years, albeit now in the guise of "poverty eradication" and "nation-building" instead of "economic growth." American foreign aid declined precipitously in the 1970s and 1980s, with the preponderance of international development dollars for the Third World flowing from such international institutions as the International Monetary Fund and the World Bank (under the leadership of former U.S. secretary of defense Robert McNamara from 1968 to 1981). In fiscal year 2001, the United States designated less than 1 percent of its federal budget ($16.8 billion) for international aid, and nearly 30 percent of this went directly to key allies Israel and Egypt in the form of military aid. The terrorist attacks of September 11, 2001, presented an opportunity for foreign policy analysts to reconsider the role of international aid in fostering global stability, but the most vocal commentators were defense hawks, who drew on old Cold War language to depict the violence as the result of a fundamental conflict between repressive Islamic fundamentalism and American freedom. It was left to the international aid community to pull Kennedy-era explanations from the Cold War handbag: radical Islamic terrorism was the result of uneven global development, a desperate cry for help from disaffected youth with no discernable access to global capital. As it had once before in Vietnam, the United States set about propelling Iraq and Afghanistan into the liberal democratic future through some combination of horrific violence and modern plumbing. The results have not been encouraging.

The notion of an American "scientific community" has always obscured the deep divisions between academic researchers, industrial scientists, and military contractors. For the brief period of the Cold War, these three groups shared a common sponsor (the federal government) and a common goal (international preeminence). But now what? Employees of entrepreneurial start-ups know their paychecks depend on identifying patentable research. Military contractors count on congressional enthusiasm for pork barrel projects and high-tech weaponry. The academic scientific community, meanwhile, has struggled to make sense of its new place in American domestic and foreign policy.

If the institutions of American science are once again to find solid foot-

*Jon Huntsman, Aug. 18, 2011, http://twitter.com/#!/JonHuntsman/status/104250677051654144, accessed Jan. 22, 2012.

ing, they must articulate some purpose that can resonate with broader national goals. One might also ask, "What cost glory?" The stories presented in this book should make clear that American science circa 1955, or even circa 1963, was neither an oasis of intellectual freedom nor an undertaking of peaceful purpose; its record of accomplishments was certainly clouded with disappointments. As Americans move into an era of globalization, they must once again ask themselves what role science and technology will play in security, prosperity, and national identity, especially now that they have become skeptical of the ability of science and technology to solve the world's problems. For if scientists and citizens fail to adequately and seriously address the question of how science and technology can help society—and, indeed, the planet—dire consequences will almost certainly follow.

Acknowledgments

Introductory works are a community effort. It would not have been possible to tell the story presented here without the tremendous outpouring of innovative scholarship on science and the Cold War that has emerged over the past twenty-five years. The bibliographic essay at the end of this volume only begins to suggest the number of books, articles, dissertations, and conference papers I voraciously consumed in the process of writing this book. I must also therefore begin by thanking my teachers, particularly Susan Lindee, Michael Katz, Robert Kohler, Charles Rosenberg, and Janet Tighe, for showing me how to navigate the scholarly literature.

My colleagues have been unusually generous in sharing their time, ideas, and opinions. Babak Ashrafi, Carin Berkowitz, Paul Burnett, Nathan Ensmenger, Darin Hayton, John Jackson, Roger Launius, Michael Katz, Patrick McCray, Suzanne Moon, Karen Rader, Joanna Radin, Corinna Schlombs, Noah Shusterman, Amy Slaton, Matt Wisnioski, and Zouyue Wang all read portions of the manuscript (some multiple times); Angela Creager, Michael Gordin, Mott Greene, Sharon Kingsland, and Bill Leslie read it in its entirety. I am also grateful for the feedback of two anonymous reviewers who read the book proposal at an early stage. The vibrant community of historians of science on Facebook—too many to mention here—offered ongoing commentary on the wisdom of including specific examples or particular turns of phrase. I alone, however, share the responsibility for whatever errors crept in along the way.

As someone who edits, as well as writes, for a living, I have had the unusual opportunity to observe a variety of different approaches to writing and finishing books. I have drawn strength from my clients' determination and tenacity during the most difficult points of writing this book. I thank them for their encouragement, enthusiasm, and patience. Bob Brugger at the Johns Hopkins University Press—a fine wordsmith himself—initially suggested the idea of this project. Kara Reiter served as the friendly face of the Press during the final stage of manuscript preparation. Julie McCarthy and Kim Johnson confidently oversaw the transition from manuscript to book, and Carolyn Moser's fine copyediting made my prose fit for publication.

Finally, I owe a special thanks to my family and friends, who are, I am sure, as delighted as I am that this project has finally reached its conclusion. Both my husband, Andrew Chalfen, and my father, Tony Wolfe, read the manuscript in its entirety and offered suggestions for improvement, with good cheer. Their encouragement and pride kept me going. Thanks, guys.

Suggested Further Reading

Introduction

My account of Project Plowshare is largely derived from Scott Kirsch, *Proving Grounds: Project Plowshare and the Unrealized Dream of Nuclear Earthmoving* (New Brunswick, NJ: Rutgers University Press, 2005). Additional information on resistance to the project can be found in Dan O'Neill, *The Firecracker Boys: H-Bombs, Inupiat Eskimos, and the Roots of the Environmental Movement* (New York: St. Martin's Press, 1994). For the government's official take on Plowshare, see U.S. Department of Energy, Nevada Operations Office, Office of Public Affairs and Information, *Project Plowshare* (Washington, DC: Government Printing Office, 2001), available online at www.osti.gov/opennet/reports/plow shar.pdf.

The full text and an audio recording of Eisenhower's farewell address is available online through the Eisenhower Presidential Library and Museum: www.eisenhower.archives.gov/research/online_documents/farewell_address .html. Additional bibliographic suggestions for most of the issues discussed in the introduction are provided in the suggested readings for individual chapters. Readers looking for an introduction to the literature on Cold War science might be best served by comparing the point of view in three works that have by now become classics: Paul Forman, "Behind Quantum Electronics: National Security as Basis for Physical Research in the United States, 1940–1960," *Historical Studies in the Physical and Biological Sciences* (*HSPS*) 18 (1987): 149–229; Daniel J. Kevles, "Cold War and Hot Physics: Science, Security, and the American State, 1945–1956," *HSPS* 20 (1990): 239–64; and Stuart W. Leslie, *The Cold War and American Science: The Military-Industrial-Academic Complex at MIT and Stanford* (New York: Columbia University Press, 1993).

For discussions of Soviet political and philosophical interventions in a number of scientific fields, see Nikolai Krementsov, *Stalinist Science* (Princeton, NJ: Princeton University Press, 1997); Ethan Pollock, *Stalin and the Soviet Science Wars* (Princeton, NJ: Princeton University Press, 2006); Loren R. Graham, *Science, Philosophy, and Human Behavior in the Soviet Union* (New York: Columbia University Press, 1987); Graham, *Science in Russia and the Soviet Union: A Short History* (Cambridge: Cambridge University Press, 1993); and Alexander Vucinich, *Empire of Knowledge: The Academy of Sciences of the USSR* (Berkeley: University of California Press, 1984).

The essays in Caroline Hannaway, ed., *Biomedicine in the Twentieth Century: Practices, Policies, and Politics* (Amsterdam: IOS Press, 2008), provide a sampling of current scholarship on the rise of American biomedicine, and how that story does and does not mesh with larger Cold War themes.

1 The Atomic Age

The literature on the Manhattan Project and nuclear weaponry is vast and contentious. Two recent books by Michael D. Gordin provide an excellent introduction to both the material itself and the scholarly literature: *Five Days in August: How World War II Became a Nuclear War* (Princeton: Princeton University Press, 2007), and *Red Cloud at Dawn: Truman, Stalin, and the End of the Atomic Monopoly* (New York: Farrar, Straus and Giroux, 2009). My discussions of the bomb's "specialness" and the importance of the Smyth Report are based on his accounts. For the Manhattan Project, see Richard Rhodes, *The Making of the Atomic Bomb* (New York: Simon and Schuster, 1986); Martin J. Sherwin, *A World Destroyed: The Atomic Bomb and the Grand Alliance* (New York: Knopf, 1975); and Daniel J. Kevles, *The Physicists: The History of a Scientific Community in Modern America*, rev. ed. (Cambridge: Harvard University Press, 1995), 324–48. J. Samuel Walker's *Prompt and Utter Destruction: Truman and the Use of Atomic Bombs against Japan*, rev. ed. (Chapel Hill: University of North Carolina Press, 2004), offers a perceptive survey of the various explanations offered for the bomb's use. For the atomic monopoly and atomic "secrets," see Gregg Herken, *The Winning Weapon: The Atomic Bomb in the Cold War, 1945–1950* (New York: Knopf, 1980). The entire Smyth Report is available online at www.atomicarchive.com/Docs/SmythReport/index.shtml. For the Soviet bomb project and its relationship to American research, see David Holloway, *Stalin and the Bomb: The Soviet Union and Atomic Energy, 1939–1956* (New Haven: Yale University Press, 1994).

The classic, if somewhat limited, account of the postwar atomic scientists' movement is Alice Kimball Smith, *A Peril and a Hope: The Scientists' Movement in America, 1945–47* (Chicago: University of Chicago Press, 1965). A more recent analysis of atomic scientists' activism, as well as the creation of the AEC and the concomitant political investigations of American scientists, can be found in Jessica Wang, *American Science in an Age of Anxiety: Scientists,*

Anticommunism, and the Cold War (Chapel Hill: University of North Carolina Press, 1999). Charles Thorpe's insightful biography of Oppenheimer also has significant background on the Manhattan Project, the atomic scientists' movement, and the debate surrounding the hydrogen bomb: *Oppenheimer: The Tragic Intellect* (Chicago: University of Chicago Press, 2006). For additional discussions of the hydrogen bomb debate and the entire text of the General Advisory Committee report, see Herbert York, *The Advisors: Oppenheimer, Teller, and the Superbomb* (San Francisco: W. H. Freeman, 1975); and Peter Galison and Barton Bernstein, "'In Any Light': Scientists and the Decision to Build the Superbomb," *HSPS* 19 (1989): 267–347.

My account of the Lawrence Livermore National Laboratory is derived from J. L. Heilbron, Robert W. Seidel, and Bruce R. Wheaton, *Lawrence and His Laboratory: Nuclear Science at Berkeley* (Berkeley: Lawrence Berkeley Laboratory and the Office for History of Science and Technology of the University of California, Berkeley, 1981).

2 The Military-Industrial Complex

Although limited to the physics community, the best introduction to the mid-twentieth-century landscape of research funding is Kevles, *The Physicists*. See also Roger L. Geiger, *Research and Relevant Knowledge: American Research Universities since World War II* (New York: Oxford University Press, 1993), and Philip Mirowski, *Science-Mart: Privatizing American Science* (Cambridge: Harvard University Press, 2011), pp. 87–138. A biography of Vannevar Bush is a useful introduction to the Office of Scientific Research and Development, the Office of Naval Research, and the establishment of the National Science Foundation: G. Pascal Zachary, *Endless Frontier: Vannevar Bush, Engineer of the American Century* (New York: Free Press, 1997). Bush's *Science: The Endless Frontier* is available online at www.nsf.gov/od/lpa/nsf50/vbush1945.htm. For the ONR, see Harvey M. Sapolsky, *Science and the Navy: The History of the Office of Naval Research* (Princeton: Princeton University Press, 1990). Discussions of the proportion of science funding supplied by the defense sector, and the meaning of that funding for scientific research, can be found in Forman, "Behind Quantum Electronics"; Kevles, "Cold War and Hot Physics"; and Leslie, *The Cold War and American Science*.

My account of the Applied Physics Laboratory relies on Michael Aaron Dennis, "'Our First Line of Defense': Two University Laboratories in the Postwar American State," *Isis* 85 (1994): 427–55, and the APL's account of its history on its website: www.jhuapl.edu/aboutapl/heritage. For Brookhaven National Laboratory and the AEC, see Peter J. Westwick, *The National Labs: Science in an American System, 1947–1974* (Cambridge: Harvard University Press, 2003); Robert W. Seidel, "A Home for Big Science: The Atomic Energy Commission's Laboratory System," *HSPS* 16 (1986): 135–75; and Brookhaven's history on its website: www.bnl.gov/bnlweb/history. My account of the System Development Corporation and its contributions to SAGE is largely derived from Martin Campbell-Kelly, *From Airline Reservations to Sonic the Hedgehog: A History of the Software Industry* (Cambridge: MIT Press, 2003), pp. 28–55; and Paul N. Edwards, *The Closed World: Computers and the Politics of Discourse in Cold War America* (Cambridge: MIT Press, 1996), pp. 75–111.

An excellent introduction to the issue of classified scientific research is Peter Galison, "Removing Knowledge," *Critical Inquiry* 31 (2004): 229–43. Alex Wellerstein's dissertation, "Knowledge and the Bomb: Nuclear Secrecy in the United States" (Ph.D. diss., Harvard University, 2010), offers a more comprehensive treatment of the topic. For additional information on the State Department's interest in scientific intelligence, see Allan A. Needell, *Science, Cold War, and the American State: Lloyd V. Berkner and the Balance of Professional Ideals* (Amsterdam: Harwood Academic Publishers in association with the National Air and Space Museum, 2000); and Ronald E. Doel, "Scientists as Policymakers, Advisors, and Intelligence Agents: Linking Contemporary Diplomatic History with the History of Contemporary Science," in Thomas Söderqvist, ed., *The Historiography of Contemporary Science and Technology* (Amsterdam: Harwood, 1997), pp. 215–44. Information on political investigations of American scientists, including Condon, can be found in Wang, *American Science in an Age of Anxiety*, and Lawrence Badash, "Science and McCarthyism," *Minerva* 38 (2000): 53–80. Oppenheimer's troubles are detailed in Thorpe, *Oppenheimer*. For a broader discussion of how anti-Communism affected the entire American academic community, see Ellen W. Schrecker, *No Ivory Tower: McCarthyism and the Universities* (New York: Oxford University Press, 1986). The implications of so much secrecy for science advising (or the lack thereof) are explored in Gregg Herken, *Cardinal Choices: Presidential Science Advising from the Atomic Bomb to SDI*, rev. and expanded ed. (Stanford: Stanford University Press, 2000), and Zouyue Wang, *In Sputnik's Shadow: The President's Science Advisory Committee and Cold War America* (New Brunswick, NJ: Rutgers University Press, 2008).

3 Big Science

Alvin Weinberg's initial diagnosis of the problems of Big Science can be found in *Science* 134 (July 21, 1961): 161–64. Readers interested in historians' take on the concept and characteristics of Big Science should start with the essays in

Peter Galison and Bruce Hevly, eds., *Big Science: The Growth of Large-Scale Research* (Stanford: Stanford University Press, 1992). Within that volume, the essays by Robert Seidel ("The Origins of the Lawrence Berkeley Laboratory," pp. 21–45) and Dominique Pestre and John Krige ("Some Thoughts on the Early History of CERN," pp. 78–99) are particularly useful in showing continuity and change in the growth of postwar research. For the big picture on Big Science, see also James H. Capshew and Karen A. Rader, "Big Science: Price to the Present," *Osiris*, 2nd ser., 7 (1992): 3–25. My discussion of E. O. Lawrence's Radiation Laboratory additionally relies on Seidel, "A Home for Big Science," and Heilbron, Seidel, and Wheaton, *Lawrence and His Laboratory*. Karen A. Rader discusses the Megamouse program at length in her *Making Mice: Standardizing Animals for American Biomedical Research, 1900–1955* (Princeton: Princeton University Press, 2004). For Glaser, Alvarez, and the bubble chambers, see Peter Galison, *Image and Logic: A Material Culture of Microphysics* (Chicago: University of Chicago Press, 1997), chap. 5. The material on MIT's Aeroelastic and Structures Laboratory comes from Leslie, *The Cold War and American Science*, pp. 87–90. For Project Revere, see Christopher Simpson, *Science of Coercion: Communication Research and Psychological Warfare, 1945– 1960* (New York: Oxford University Press, 1994), p. 76. Similar stories about the effects of defense funding on scientific research practices could be told for any number of disciplines; the case of the earth sciences has been particularly well told in essays by Ronald E. Doel, Kristine C. Harper, Naomi Oreskes, Kai-Henrik Barth, and Allison MacFarlane in a special issue of *Social Studies of Science* 33 (2003), ed. John Cloud.

The authoritative account of the changing postwar university is Geiger, *Research and Relevant Knowledge*; data on the number of American Ph.D.'s comes from page 217. Rebecca S. Lowen, *Creating the Cold War University: The Transformation of Stanford* (Berkeley: University of California Press, 1997), and Leslie, *The Cold War and American Science*, provide detailed accounts of the growth of two specific universities, Stanford and MIT. Leslie, in particular, discusses the ways that military agencies relied on the universities to provide technical training for their officers.

For my discussion of scientific training as a "manpower" problem, I am indebted to David Kaiser, "Cold War Requisitions, Scientific Manpower, and the Production of American Physicists after World War II," *HSPS* 33 (2002): 131–59. The statistics on physics Ph.D.'s also derives from his account. Draper's pedagogical aspirations are discussed in Leslie, *The Cold War and American Science*, pp. 90–100; Dennis, " 'Our First Line of Defense' "; and Donald MacKenzie, *Inventing Accuracy: An Historical Sociology of Nuclear Missile Guidance*

(Cambridge: MIT Press, 1990). For *Sputnik*-related curriculum reforms, particularly the Physical Science Study Committee, see John L. Rudolph, *Scientists in the Classroom: The Cold War Reconstruction of American Science Education* (New York: Palgrave, 2002). The broader context of these reforms and background on the National Defense Education Act is provided by Robert A. Divine, *The Sputnik Challenge* (New York: Oxford University Press, 1993), pp. 89–93, and Barbara Barksdale Clowse, *Brainpower for the Cold War: The Sputnik Crisis and the National Defense Education Act of 1958* (Westport, CT: Greenwood Press, 1981). Nathan Ensmenger's *The Computer Boys Take Over: Computers, Programmers, and the Politics of Technical Expertise* (Cambridge: MIT Press, 2010) contains a fascinating discussion of attempts to envision the perfect programmer. An essential resource on employment patterns for women and, to a lesser extent, minorities is Margaret W. Rossiter, *Women Scientists in America: Before Affirmative Action, 1940–1972* (Baltimore: Johns Hopkins University Press, 1995), particularly chap. 3, "'Scientific Womanpower.'"

For a discussion of attempts to establish a federal scientific advisory apparatus, see Herken, *Cardinal Choices*, and Z. Wang, *In Sputnik's Shadow*. The Seaborg Report and its implications for the expansion of the research university are discussed in Geiger, *Research and Relevant Knowledge*, pp. 166–79. Industrial research practices are discussed by Steven Shapin, *The Scientific Life: A Moral History of a Late Modern Vocation* (Chicago: University of Chicago Press, 2008), chap. 5, "Who Is the Industrial Scientist? The View from the Managers." Derek J. de Solla Price's discussion of exponential growth of science in general appears throughout *Science Since Babylon* (New Haven: Yale University Press, 1961); the growth of American science goes under the microscope in pages 1–32 of his *Big Science, Little Science* (New York: Columbia University Press, 1963). Kennedy's "hearts and minds" speech was delivered to the United Nations on September 20, 1963; full text and audio are among the selected speeches available at the John F. Kennedy Presidential Library website, www.jfklibrary.org/Research/Ready-Reference.

4 Hearts and Minds and Markets

The idea that the battle for Third World hearts and minds should be understood primarily as an ideological one has been advanced most forcefully and convincingly by Odd Arne Westad in *The Global Cold War: Third World Interventions and the Making of Our Times* (New York: Cambridge University Press, 2005). John Lewis Gaddis tells a shorter, and more European, version of this story in his *The Cold War: A New History* (New York: Penguin, 2005). Both books are essential to understanding the global sweep of the Cold War conflict.

For the role of scientific aid in the Marshall Plan, see John Krige, *American Hegemony and the Postwar Reconstruction of Science in Europe* (Cambridge: MIT Press, 2006). John DiMoia describes attempts to rebuild South Korea's electrical power network in "Atoms for Sale? Cold War Institution-Building and the South Korean Atomic Energy Project, 1945–1965," *Technology and Culture* 51 (2010): 589–618; David Ekbladh describes the Korean development efforts as part of the larger sweep of American modernization theory in *The Great American Mission: Modernization and the Construction of an American World Order* (Princeton: Princeton University Press, 2010), pp. 114–52. For the broader context of what might be called scientific diplomacy, see Kenneth Osgood, *Total Cold War: Eisenhower's Secret Propaganda Battle at Home and Abroad* (Lawrence: University Press of Kansas, 2006). Truman's comments on the Point Four program are from his 1949 Inaugural Address; the text and audio of the speech can be found online at the Harry S. Truman Presidential Library and Museum website, www.trumanlibrary.org/educ/inaug.htm.

In addition to Ekbladh, *The Great American Mission,* two useful introductions to American social scientific ideas about international development are Michael E. Latham, *Modernization as Ideology: American Social Science and 'Nation Building' in the Kennedy Era* (Chapel Hill: University of North Carolina Press, 2000), and Nils Gilman, *Mandarins of the Future: Modernization Theory in Cold War America* (Baltimore: Johns Hopkins University Press, 2003). Latham discusses the Peace Corps, the Alliance for Progress, and the Strategic Hamlet Program; Gilman covers the Center for International Studies at MIT. Nick Cullather's *The Hungry World: America's Cold War Battle against Poverty in Asia* (Cambridge: Harvard University Press, 2010) extends the history of U.S. development work to the 1920s and offers a cogent analysis of the Green Revolution. An additional perspective on U.S. public and private efforts to influence global agriculture is John H. Perkins, *Geopolitics and the Green Revolution: Wheat, Genes, and the Cold War* (New York: Oxford University Press, 1997).

The classic text on postwar American development theory is W. W. Rostow, *The Stages of Economic Growth: A Non-Communist Manifesto* (Cambridge: Cambridge University Press, 1960). All Rostow quotations are taken from this edition. For Rostow's biography, see David Milne, *America's Rasputin: Walt Rostow and the Vietnam War* (New York: Hill and Wang, 2008). An influential interpretation of how economists came to use GDP as a measure of growth is provided by Arturo Escobar in *Encountering Development: The Making and Unmaking of the Third World* (Princeton: Princeton University Press, 1995). Though written from the perspective of a practicing economist, William Easterly's *The Elusive Quest for Growth: Economists' Adventures and Misadventures in the Tropics* (Cambridge: MIT Press, 2001) nevertheless provides a useful sketch of the history of development economics; the statistics on U.S. foreign aid spending are his. USAID provides detailed information on foreign aid in its series of "Greenbook" publications. The most recent is *U.S. Overseas Loans and Grants: Obligations and Loan Authorizations, July 1, 1945–September 30, 2010*, available online at http://gbk.eads.usaidallnet.gov/docs.

The historical scholarship on science- and technology-focused international development projects is expanding rapidly. The essays in the following volumes are particularly useful introductions to the scale, scope, and variety of these projects: Frederick Cooper and Randall Packard, eds., *International Development and the Social Sciences: Essays on the History and Politics of Knowledge* (Berkeley: University of California Press, 1997); John Krige and Kai-Henrik Barth, eds., "Global Power Knowledge: Science and Technology in International Affairs," *Osiris*, 2nd ser., 21 (2006); and Carol E. Harrison and Ann Johnson, eds., "National Identity: The Role of Science and Technology," *Osiris*, 2nd ser., 24 (2009).

For the specific examples discussed in this chapter, see Helen Tilley, *Africa as a Living Laboratory: Empire, Development, and the Problem of Scientific Knowledge, 1870–1950* (Chicago: University of Chicago Press, 2011), pp. 115–68; Heather J. Hoag and May-Britt Öhman, "Turning Water into Power: Debates over the Development of Tanzania's Rufiji River Basin, 1945–1985," *Technology and Culture* 49 (2008): 624–51; Fumbuka Ng'wanakilala, "Tanzania Plans $2 Billion Hydro Plant with Brazil," *Reuters Africa*, Dec. 1, 2010, http://af.reuters .com/article/topNews/idAFJOE6B00612010I201, accessed July 19, 2011; Stuart W. Leslie and Robert Kargon, "Exporting MIT: Science, Technology, and Nation-Building in India and Iran," *Osiris*, 2nd ser., 21 (2006): 110–30; Ross Bassett, "Aligning India in the Cold War Era: Indian Technical Elites, the Indian Institute of Technology at Kanpur, and Computing in India and the United States," *Technology and Culture* 50 (2009): 783–810; Randall M. Packard, *The Making of a Tropical Disease: A Short History of Malaria* (Baltimore: Johns Hopkins University Press, 2007); and Marcos Cueto, *Cold War, Deadly Fevers: Malaria Eradication in Mexico, 1955–1975* (Washington, DC, and Baltimore: Woodrow Wilson Center Press and Johns Hopkins University Press, 2007). For an astute analysis of U.S. attempts to develop the Mekong River Delta even while escalating the bombing in North Vietnam, see Ekbladh, *The Great American Mission*, chap. 6, "A TVA on the Mekong."

Finally, no account of grand development schemes, and how they failed, is complete without mention of James C. Scott's *Seeing Like a State: How*

Certain Schemes to Improve the Human Condition Have Failed (New Haven: Yale University Press, 1998). Scott's notion of "high modernism" has become a critical tool in making sense of how modern states use technology to render their societies legible, and thus subject to control. Chapter 7, "Compulsory Villagization in Tanzania" (pp. 223–61), specifically discusses Nyerere's plans in East Africa, including his infatuation with the TVA.

5 Science and the General Welfare

The best introduction to postwar social scientific understandings of affluence, race, and the crisis of poverty is through the classic texts themselves, most of which are surprisingly readable (if often long). Readers are therefore encouraged to turn directly to John Kenneth Galbraith, *The Affluent Society* (Boston: Houghton Mifflin, 1958); Michael Harrington, *The Other America: Poverty in the United States* (New York: Macmillan, 1962); Gunnar Myrdal with Richard Sterner and Arnold Rose, *An American Dilemma: The Negro Problem and Modern Democracy* (New York: Harper, 1944); E. Franklin Frazier, *The Negro Family in the United States* (Chicago: University of Chicago Press, 1939); and Office of Policy Planning and Research, U.S. Department of Labor, *The Negro Family: The Case for National Action* (Washington, DC: Government Printing Office, 1965).

For an excellent introduction to the economic and cultural phenomenon of postwar consumerism, see Lizabeth Cohen, *A Consumer's Republic: The Politics of Mass Consumption in Postwar America* (New York: Knopf, 2003). The statistics on family income and consumer spending are from page 121 of her account. For background on federal investment in science and technology see (besides the references provided for chapters 2 and 3), National Science Foundation, Division of Science Resource Studies, *Federal Funds for Research and Development: Detailed Historical Tables, Fiscal Years 1951–2001* (Arlington, VA: NSF, 2001), available online as www.nsf.gov/statistics/nsf01334. The quotations from Vannevar Bush are from the summary portion of *Science: The Endless Frontier*. I am grateful to Matthew Wisnioski for bringing the Engineers Joint Council report to my attention; the full citation is Engineers Joint Council, Engineering Research Committee, *The Nation's Engineering Research Needs, 1965–1985: Summary Report*, May 25, 1962. Wisnioski discusses the engineering community's concerns about research directions in *Engineers for Change: Competing Visions of Technology in 1960s America* (Cambridge: MIT Press, 2012). The story of the divergent fates of the American and Japanese semiconductor industries is told in Hyungsub Choi, "Manufacturing Knowledge in Transit: Technical Practices, Organizational Change, and the Rise of the Semiconductor Industry in the United States and Japan, 1948–1960" (Ph.D. diss., Johns Hopkins University, 2007). Margaret Pugh O'Mara's *Cities of Knowledge: Cold War Science and the Search for the Next Silicon Valley* (Princeton: Princeton University Press, 2005), provides a fascinating look at attempts to use investments in science and technology as the basis for regional economic prosperity.

The relationship of the Civil Rights movement to Cold War politics is discussed in Brenda Gayle Plummer, *Rising Wind: Black Americans and U.S. Foreign Affairs, 1935–1960* (Chapel Hill: University of North Carolina Press, 1996); Mary L. Dudziak, *Cold War Civil Rights: Race and the Image of American Democracy* (Princeton: Princeton University Press, 2000); and Laura A. Belmonte, *Selling the American Way: U.S. Propaganda and the Cold War* (Philadelphia: University of Pennsylvania Press, 2008), pp. 159–77.

For a cultural and political history of postwar psychology, see Ellen Herman, *The Romance of American Psychology: Political Culture in the Age of Experts* (Berkeley: University of California Press, 1995); a version of Herman's chapter 5, "The Career of Cold War Psychology," also appears under the same title in *Radical History Review* 63 (1995): 52–85. Additional information on the funding structure of the postwar social sciences, including psychology, can be found in Hunter Crowther-Heyck, "Patrons of the Revolution: Ideas and Institutions in Postwar Behavioral Sciences," *Isis* 97 (2006): 420–46; and Joy Rohde, "Gray Matters: Social Scientists, Military Patronage, and Democracy in the Cold War," *Journal of American History* 96 (2009): 99–122. Joel Issac, "The Human Sciences in Cold War America," *The Historical Journal* 50 (2007): 725–46, provides a useful corrective to a literature that perhaps gives too much stress to direct military influence and not enough to broader cultural concerns.

For psychological theories of race, see (in addition to Herman, *Romance*) Alice O'Conner, *Poverty Knowledge: Social Science, Social Policy, and the Poor in Twentieth-Century U.S. History* (Princeton: Princeton University Press, 2001); and John P. Jackson, Jr., *Social Scientists for Social Justice: Making the Case against Segregation* (New York: New York University Press, 2001). Walter A. Jackson's biography of Gunnar Myrdal, *Gunnar Myrdal and America's Conscience: Social Engineering and Racial Liberalism, 1938–1987* (Chapel Hill: University of North Carolina Press, 1990), also contains useful background information on this topic. The context and consequences of the Moynihan Report are discussed in James T. Patterson, *Freedom Is Not Enough: The Moynihan Report and America's Struggle over Black Family Life from LBJ to Obama* (New York: Basic Books, 2010).

A useful discussion of the Great Society programs in the context of larger American political movements is Gareth Davies, *From Opportunity to Entitle-*

ment: The Transformation and Decline of Great Society Liberalism (Lawrence: University Press of Kansas, 1996). For the role of social scientific knowledge in designing and operating the programs, see O'Conner, *Poverty Knowledge*, and Michael A. Bernstein, *A Perilous Progress: Economists and Public Purpose in Twentieth-Century America* (Princeton: Princeton University Press, 2001). The literature on operations research and systems analysis is (perhaps not surprisingly) mostly technical, but a good historical overview is the editors' introductory essay in Agatha C. Hughes and Thomas P. Hughes, eds., *Systems, Experts, and Computers: The Systems Approach in Management and Engineering, World War II and After* (Cambridge: MIT Press, 2000), pp. 1–26.

For case studies of specific attempts to apply defense strategies to the problems of the American city, see two chapters in Hughes and Hughes—David R. Jardini, "Out of the Blue Yonder: The Transfer of Systems Thinking from the Pentagon to the Great Society, 1961–1965" (pp. 311–58), and Davis Dyer, "The Limits of Technology Transfer: Civil Systems at TRW, 1965–1975" (pp. 359–84)—as well as Jennifer S. Light, *From Warfare to Welfare: Defense Intellectuals and Urban Problems in Cold War America* (Baltimore: Johns Hopkins University Press, 2003).

6 The Race to the Moon

An enormous number of books have been written about Apollo and the space race; surprisingly few of them attempt to place the story within the larger frame of either the Cold War or postwar American history. Readers hoping to orient themselves in this literature are advised to start with several excellent historiographical essays. Some of the best include Roger D. Launius, "Interpreting the Moon Landings: Project Apollo and the Historians," *History and Technology* 22 (2006): 225–55; Launius, "The Historical Dimension of Space Exploration: Reflections and Possibilities," *Space Policy* 16 (2000): 23–38; Asif A. Siddiqi, "Competing Technologies, National(ist) Narratives, and Universal Claims: Toward a Global History of Space Exploration," *Technology and Culture* 51 (2010): 425–43; and the essays in Steven J. Dick and Roger D. Launius, eds., *Critical Issues in the History of Spaceflight*, NASA SP-2006-4702 (Washington, DC: Government Printing Office, 2006), and Steven J. Dick, ed., *Remembering the Space Age: Proceedings of the Fiftieth Anniversary Conference*, NASA SP-2008-4703 (Washington, DC: Government Printing Office, 2008). All NASA publications mentioned here are available online through the NASA History Program Office website, http://history.nasa.gov.

The most accessible overall introduction to the Apollo program is Charles Murray and Catherine Bly Cox, *Apollo: The Race to the Moon* (New York: Simon and Schuster, 1989). For a journalistic account focusing on the astronauts, see Andrew Chaikin, *A Man on the Moon: The Voyages of the Apollo Astronauts* (New York: Viking, 1994). Astronaut Michael Collins's fascinating memoir, *Carrying the Fire: An Astronaut's Journeys* (New York: Farrar, Straus and Giroux, 1974), doubles as a history of manned spaceflight. For a technical overview of Apollo, including mission-by-mission summaries, see William David Compton, *Where No Man Has Gone Before: A History of Apollo Lunar Exploration Missions*, NASA SP-4214 (Washington, DC: Government Printing Office, 1989).

The classic political histories of the space race are Walter A. McDougall, *. . . The Heavens and the Earth: A Political History of the Space Age* (New York: Basic Books, 1985), and John M. Logsdon, *John F. Kennedy and the Race to the Moon* (New York: Palgrave Macmillan, 2010). (Logsdon's 2010 work should be considered a replacement for his earlier work of a similar title.) Though dated, Vernon Van Dyke's *Pride and Power: The Rationale of the Space Program* (Urbana: University of Illinois Press, 1964) provides a useful schematic of possible justifications for a lunar mission before the final outcome was known. William E. Burrows's *This New Ocean: The Story of the First Space Age* (New York: Random House, 1998) contains a thorough and lucid discussion of the space race through the lens of U.S.-Soviet competition. For public support of Apollo (or lack thereof), see Roger D. Launius, "Public Opinion Polls and Perceptions of U.S. Human Spaceflight," *Space Policy* 19 (2003): 163–75.

The above accounts provide reasonable introductions to most of the topics discussed in this chapter. For additional information on 1950s- and 1960s-era space defense, see David H. DeVorkin, *Science with a Vengeance: How the Military Created the U.S. Space Sciences after World War II* (New York: Springer-Verlag, 1992); William E. Burrows, *Deep Black: Space Espionage and National Security* (New York: Random House, 1986); and Curtis Peebles, *The CORONA Project: America's First Spy Satellites* (Annapolis: Naval Institute Press, 1997). Michael J. Neufeld's biography of Wernher von Braun, *Von Braun: Dreamer of Space, Engineer of War* (New York: Knopf, 2007), is an important corrective to a literature that has too often been willing to skip over the German rocket scientist's Nazi past.

Little has been written about NASA's relationship to the Great Society—a gap that is surprising, given Lyndon Johnson's championing of the space race as an economic development project for the American South and West. Three exceptions are W. Henry Lambright's *Powering Apollo: James E. Webb of NASA* (Baltimore: Johns Hopkins University Press, 1995); Robert Dallek, "Johnson,

Project Apollo, and the Politics of Space Program Planning," in Roger D. Launius and Howard E. McCurdy, eds., *Spaceflight and the Myth of Presidential
Leadership* (Urbana: University of Illinois Press, 1997), pp. 68–91; and Roger D.
Launius, "Managing the Unmanageable: Apollo, Space Age Management, and
American Social Problems," *Space Policy* 24 (2008): 158–65. See also James E.
Webb's *Space Age Management: The Large-Scale Approach* (New York: McGraw
Hill, 1969). The book itself is dreadfully dull, but its mere existence is fascinating. For excellent discussions of the engineering and managerial challenges of
Apollo, see Stephen B. Johnson, *The Secret of Apollo: Systems Management in
American and European Space Programs* (Baltimore: Johns Hopkins University
Press, 2002). The phrase "bureaucracy of innovation" is from Johnson, *The
Secret of Apollo*. Man-and-machine issues are discussed in Collins, *Carrying the
Fire*, and David A. Mindell, *Digital Apollo: Human and Machine in Spaceflight*
(Cambridge: MIT Press, 2008).

For Kennedy's dreams of a bilateral lunar mission, see Logsdon, *John F.
Kennedy*, and Matthew J. Von Bencke, *The Politics of Space: A History of U.S.-
Soviet/Russian Competition and Cooperation in Space* (Boulder, CO: Westview
Press, 1997). McDougall discusses what he calls the "benign hypocrisy" of
American attempts to foster internationalism at length in . . . *The Heavens and
the Earth*. For the fascinating story of the Moonwatch program and amateur
contributions to the International Geophysical Year, see W. Patrick McCray,
*Keep Watching the Skies! The Story of Operation Moonwatch and the Dawn of
the Space Age* (Princeton: Princeton University Press, 2008). The political and
economic implications of the United States' role in creating the first global
communications satellite network are discussed in Hugh R. Slotten, "Satellite
Communications, Globalization, and the Cold War," *Technology and Culture*
43 (2002): 315–50. John Krige tells the story of aborted American assistance to
the European Launcher Development Organization in "Technology, Foreign
Policy, and International Cooperation in Space," in Dick and Launius, *Critical Issues*, pp. 239–60. For the Apollo-Soyuz Test Project, see Von Bencke, *The
Politics of Space*, and Edward Clinton Ezell and Linda Neuman Ezell, *The
Partnership: A History of the Apollo-Soyuz Test Project*, NASA SP-4209 (Washington, DC, 1978). On Soviet antisatellite technologies, see Paul B. Stares, *The
Militarization of Space: U.S. Policy, 1945–1984* (Ithaca, NY: Cornell University
Press, 1985).

Finally, my discussion of *Earthrise* is indebted to Denis Cosgrove, "Contested Global Visions: *One-World, Whole-Earth*, and the Apollo Space Photographs," *Annals of the Association of American Geographers* 84 (1994): 270–94, and Robert Poole, *Earthrise: How Man First Saw the Earth* (New Haven: Yale University Press, 2008).

7 The End of Consensus

First things first. Yes, some scientists were hippies. For two fascinating case studies, see David Kaiser, *How the Hippies Saved Physics: Science, Counterculture, and the Quantum Revival* (New York: Norton, 2011), and Fred Turner, *From Counterculture to Cyberculture: Stewart Brand, the Whole Earth Network, and the Rise of Digital Utopianism* (Chicago: University of Chicago Press, 2006).

Though limited to the Eisenhower years, the best introduction to the debate about nuclear fallout remains Robert A. Divine's *Blowing on the Wind: The Nuclear Test Ban Debate, 1954–1960* (New York: Oxford University Press, 1978). A graphical depiction of the history of the Doomsday Clock can be seen on the website of the *Bulletin of Atomic Scientists*, www.thebulletin.org/content/doomsday-clock/timeline. The complete archives of the *Bulletin* are available publicly through Google Books, http://books.google.com.

The attempts of American geneticists to come to terms with the biological effects of atomic radiation on human heredity are treated in M. Susan Lindee, *Suffering Made Real: American Science and the Survivors at Hiroshima* (Chicago: University of Chicago Press, 1994); John Beatty, "Weighing the Risks: Stalemate in the Classical/Balance Controversy," *Journal of the History of Biology* (*JHB*) 20 (1987): 289–319; and Jacob Darwin Hamblin, "'A Dispassionate and Objective Effort': Negotiating the First Study on the Biological Effects of Atomic Radiation," *JHB* 40 (2007): 147–77. Bentley Glass's activism is little known but well documented in his archival collection at the American Philosophical Society Library in Philadelphia. For a preliminary survey of this collection and a sketch of Glass's life, see Audra J. Wolfe, "The Organization Man and the Archive: A Look at the Bentley Glass Papers," *JHB* 44 (2011): 147–51. Many documents relating to Linus Pauling's political activism—including the text of his nuclear test ban petition, signature forms, and the Senate Internal Security Subcommittee subpoena—are available at the Oregon State University Libraries Special Collections website, http://osulibrary.oregonstate.edu/specialcollections/coll/pauling/peace/papers. For the Committee on Nuclear Information, see Kelly Moore, *Disrupting Science: Social Movements, American Scientists, and the Politics of the Military, 1945–1975* (Princeton: Princeton University Press, 2008), pp. 96–129; and Michael Egan, *Barry Commoner and the Science of Survival: The Remaking of American Environmentalism* (Cambridge: MIT Press, 2007), pp. 66–72.

Geiger's *Research and Relevant Knowledge*, pp. 230–69, is an excellent survey of the effects of campus protests on the politics of research funding, including the Mansfield Amendment. Another good introduction to the events at MIT, including the March 4 protests and Ascher Shapiro's views, is Leslie's *The Cold War and American Science*, pp. 233–56. For Project Camelot, see Ellen Herman, "Project Camelot and the Career of Cold War Psychology," in Christopher Simpson, ed., *Universities and Empire: Money and Politics in the Social Sciences during the Cold War* (New York: New Press, 1998), 97–133. Joy Rohde provides additional context for the Special Operations Research Organization in "Gray Matters." Matt Wisnioski examines the response of both mainstream and radical engineering students and faculty to defense work at Princeton, MIT, and several other campuses in "Inside 'the System': Engineers, Scientists, and the Boundaries of Social Protest in the Long 1960s," *History and Technology* 19 (2003): 313–33. Moore discusses a wide variety of radical science groups in *Disrupting Science*, pp. 158–89; the group's allegations against Seaborg are detailed on page 167.

Though not a scholarly account, Ann Finkbeiner's *The Jasons: The Secret History of Science's Postwar Elite* (New York: Viking, 2006) is the most thorough source available on the group; see also the Scientists and Engineers for Social and Political Action booklet *Science against the People: The Story of Jason* (Berkeley, CA: SESPA, 1972). For the ripple effects on the divisions within the Jasons throughout the federal science advising community, see Herken, *Cardinal Choices*, pp. 152–56. Kistiakowsky's resignation was originally reported in Daniel Greenberg, "Kistiakowsky Cuts Defense Department Ties over Vietnam," *Science* 159 (1 March 1968): 958, and is discussed in Z. Wang, *In Sputnik's Shadow*, pp. 260–61. My discussion of Nixon's decision to dissolve the President's Science Advisory Committee relies on Herken's and Z. Wang's accounts.

8 Cold War Redux

For a political overview of the end of the Cold War, see Gaddis, *The Cold War*, 197–257. The effects of the 1980s-era military buildup on research universities are discussed, in a general sense, in Geiger, *Research and Relevant Knowledge*, 310–37, and Leslie, *The Cold War and American Science*, 249–56. Mirowski's *Science-Mart* is an important new work that puts forth the contentious argument that the discussion of Cold War funding patterns is a distraction from the larger historical shift toward neoliberalism, including the idea that all knowledge is subject to evaluation by the market. I have found Thomas Borstelmann, *The 1970s: A New Global History from Civil Rights to Economic Inequality* (Princeton: Princeton University Press, 2012), and Daniel T. Rodgers, *Age of Fracture* (Cambridge: The Belknap Press of Harvard University

Press, 2011), particularly useful in coming to grips with the social, economic, and intellectual currents that rocked American culture in the 1970s and 1980s.

For the specific data on defense R&D as a percentage of GDP, see table 3.1, "Outlays by Superfunction and Function, 1940–2016," in Office of Management and Budget, "FY2012 Historical Tables," available online at www.white house.gov/sites/default/files/omb/budget/fy2012/assets/hist.pdf, and the National Science Foundation, "Federal Obligations for Total Research and Development, by Major Agency and Performer: Fiscal Years 1951–2001," available online at www.nsf.gov/statistics/nsf01334/pdf/histb.pdf. Data on the trade deficit are available from the U.S. Census Bureau at www.census.gov/foreign -trade/statistics/historical.

The most thorough account of the origins of the biotechnology industry is Susan Wright, *Molecular Politics: Developing American and British Regulatory Policy for Genetic Engineering, 1972–1982* (Chicago: University of Chicago Press, 1994). See also Sally Smith Hughes, *Genentech: The Beginnings of Biotech* (Chicago: University of Chicago Press, 2011); Martin Kenney, "Biotechnology and the Creation of a New Economic Space," in Arnold Thackray, ed., *Private Science: Biotechnology and the Rise of the Molecular Sciences* (Philadelphia: University of Pennsylvania Press, 1998), pp. 131–43; and Doogab Yi, "Who Owns What? Private Ownership and the Public Interest in Recombinant DNA Technology in the 1970s," *Isis* 102 (2011): 446–74. For the more general context of commercialization within the university, see Mirowski, *Science-Mart*; Sheila Slaughter and Larry L. Leslie, *Academic Capitalism: Politics, Policies, and the Entrepreneurial University* (Baltimore: Johns Hopkins University Press, 1997); and Sheila Slaughter and Gary Rhoades, *Academic Capitalism and the New Economy* (Baltimore: Johns Hopkins University Press, 2004).

The most concise introduction to the politics of the Strategic Defense Initiative is also the most recent: Paul S. Boyer, "Selling Star Wars: Ronald Reagan's Strategic Defense Initiative," in Kenneth Osgood and Andrew K. Frank, eds., *Selling War in a Media Age: The Presidency and Public Opinion in the American Century* (Gainesville: University Press of Florida, 2010), pp. 196–223. Equally useful, though much longer, is Frances FitzGerald, *Way Out There in the Blue: Reagan, Star Wars, and the End of the Cold War* (New York: Simon and Schuster, 2000). For SDI in the context of science advising, see Herken, *Cardinal Choices*, pp. 199–216. A blow-by-blow account of the actions of the High Frontier Panel can be found in Donald R. Baucom, *The Origins of SDI, 1944–1983* (Lawrence: University Press of Kansas, 1992), pp. 141–70. The broader context of California-based efforts to promote private investment in space is discussed in W. Patrick McCray, "From L5 to X Prize: California's Alternative Space Movement," in Peter J. Westwick and William Deverell, eds., *Blue Sky Metropolis: The Aerospace Century in Southern California* (Berkeley: University of California Press, 2012), pp. 171–93. Joan Lisa Bromberg discusses Reagan's attitudes toward commercialization of space in *NASA and the Space Industry* (Baltimore: Johns Hopkins University Press, 1999), pp. 114–48. Specific data on SDI funding, including information on defense contractors and their political contributions, may be found in Erik Pratt, John Pike, and Daniel Lindley, "SDI Contracting: Building a Star Wars Constituency," in Gerald M. Steinberg, ed., *Lost in Space: The Domestic Politics of the Strategic Defense Initiative* (Lexington, MA: Lexington Books, 1988), pp. 111–44. Finally, for Soviet responses to SDI, see Mira Duric, *The Strategic Defence Initiative: US Policy and the Soviet Union* (Burlington, VT: Ashgate, 2003).

Epilogue

Although the authors do not quite share my interpretation, my view on the relationship between the collapse of Communism and the rise of skepticism concerning the utility of science has been greatly influenced by the stories told in Naomi Oreskes and Erik M. Conway, *Merchants of Doubt: How a Handful of Scientists Obscured the Truth on Issues from Tobacco Smoke to Global Warming* (New York: Bloomsbury Press, 2010). John Holdren's AAAS lecture is recounted from my personal experience; for the partisan divide in science during the George W. Bush years, see Chris Mooney, *The Republican War on Science* (New York: Basic Books, 2005). A much ballyhooed Pew Research Center Study found that only 6 percent of AAAS members self-identify as Republicans: Pew Research Center, "Public Praises Science; Scientists Fault Public, Media," 9 July 2009, www.people-press.org/2009/07/09/public-praises -science-scientists-fault-public-media.

Index

Page numbers in *italics* refer to illustrations.

Department of Defense; military-industrial
complex; state
graduate education, 46
Graham, Daniel O., 129
Great Society programs, 74–75, 82, 83–87, 98
Green Revolution, 4, 58
Groves, Leslie, 14–15, 24, 25, 31

Harrington, Michael, 78, 88
Harvard Medical School, 126
Head Start, 85–86
Heritage Foundation, 129
High Frontier Panel, 129–31
Holdren, John, 137–38
House Committee on Un-American
Activities, 35
human capital, concept of, 84
Human Genome Project, 137
Huntsman, Jon, 137–38
hydrogen bombs, 17–21, 37

ICBMs, 90, 128, 130
ideology in Cold War, 55–60
IIT Kanpur, 68–70
India, 4, 58, 68–70
individuals, social sciences focus on, 79–81,
83, 88
Institute for Defense Analysis, 112, 119
instrumentalism, 41–42, 44
Instrumentation Laboratory, 46–47, 112–13
intellectual property, 125, 126–28
international cooperation in space, 99–103
International Geophysical Year, 100
International Telecommunications Satellite
Consortium, 101
Iron Triangle, 132

Japan, 9, 77–78, 107, 123
Jasons, 116–17, 118, 129
Johns Hopkins University, Applied Physics
Laboratory, 28–29, 114
Johnson, Lyndon: Moynihan Report and,
81, 82–83; Rostow and, 62; social scientists
and, 83; space race and, 93, 94, 96; "War
on Poverty," 74, 86–87. See also Great Society programs

Kelkar, P. K., 69
Kennedy, John F.: Partial Test Ban Treaty, 92,
110; on path to peace, 53; Peace Corps, 64;

Rostow and, 62; social scientists and, 60;
space race and, 93–94, 95, 99
Kershaw, Joseph, 84–85
Keyworth, George, 129
Khrushchev, Nikita, 53, 57, 92, 99, 110
Kilgore, Harley, 24, 25, 38
Killian, James, 37, 48–49, 50
Kistiakowsky, George, 51, 118
Kuboyama, Aikichi, 107
Kurchatov, Igor, 21

Latin America, Alliance for Progress Program
in, 63–64
Lawrence, E. O., 17, 19, 30–31, 41, 48
Lawrence Berkeley National Laboratory:
bubble chamber, 44; fission research, 17;
origins as Radiation Laboratory, 31, 41, 43
Lilienthal, David, 16
linear programming, 84, 87
Livermore National Laboratory, 19, 20, 30
Los Alamos National Laboratory, 17, 20, 30,
34, 137

Magnuson, Warren, 24
malaria, efforts to eradicate, 70–72
management: of Apollo program, 96–99; of
Big Science, 48–53
Manhattan Project: overview, 10–13, 24, 34,
36; sites, 11–13, 12, 16
manpower for science, 45–48
Mansfield Amendment, 114, 119, 122–23
Mao Zedong, 57, 122
Marshall Plan, 56, 75
Massachusetts Institute of Technology
(MIT): Aeroelastic and Structures Labora-
tory, 44; Center for International Studies,
61, 68; DuBridge and, 18; federal contract
research laboratory of, 29; Fluid Mechan-
ics Laboratory, 113, 114; IIT Kanpur and,
68–69; Instrumentation Laboratory,
46–47; Lincoln Laboratory, 31–32, 113;
March 4 protests and, 112
matriarchal family thesis, 81
May-Johnson Bill, 15, 16
McCarthy, Joseph, 106
McMahon Bill, 15
McNamara, Robert, 83–84, 94, 116, 118, 138
Mees, C. E. Kenneth, 52
Mexico, malaria eradication programs in,
70–72